ADVANCED PLACEMENT EXAMINATION IN BIOLOGY

ADVANCED PLACEMENT EXAMINATION IN BIOLOGY

RICHARD F. HELLER, PH.D.

Professor, Department of Biology and Medical Laboratory Technology
Bronx Community College of the City University of New York
Professor of Medicine, Mount Sinai School of Medicine, New York, New York

RACHAEL F. HELLER, PH.D.

Assistant Clinical Professor of Medicine,
Mount Sinai School of Medicine, New York, New York

MACMILLAN • USA

Second Edition

Macmillan General Reference
A Prentice Hall Macmillan Company
15 Columbus Circle
New York, NY 10023

An Arco Book

Library of Congress Cataloging-in-Publication Data

Heller, Richard F.
Advanced placement examination in biology/Richard F. Heller, Rachael F. Heller.
p. cm.
ISBN 0-671-86582-X
1. Biology—Examinations, questions, etc. I. Heller, Rachael F.
QH316.H45 1990
574.076—dc 20 90-30199
CIP

Manufactured in the United States of America

15 14 13 12 11 10 9 8 7 6

TO THE MEMORY OF
MILDRED MARENGO,
RICHARD'S FIRST BIOLOGY TEACHER
AND ROLE MODEL

ACKNOWLEDGMENTS

Thanks to the Educational Testing Service for providing the information necessary to plan this revision of our book.

Also, thanks to Ted and Michele and Roger. Without their help this book would not have been possible.

TABLE OF CONTENTS

PART TWO
DIAGNOSTIC MODEL EXAMINATION

PART THREE
IMPORTANT TOPICS FOR REVIEW

PART FOUR
FINAL MODEL EXAMINATIONS

APPENDIX

PREFACE

This review book is designed for high-school students who have had some experience in biology and wish to prepare for the Advanced Placement Examination (AP) in Biology. The text includes summaries of important topics in biology, strategies to be used in preparing for and taking the examination, and additional materials that will prove useful in earning advanced standing or college credit for biology knowledge acquired in high school. Each topic in the review begins with a list of words to be defined and concludes with a miniquiz with answers and explanations. The model examinations in this book have been carefully constructed to reflect as closely as possible the actual AP examination in biology and to provide a realistic assessment of a student's readiness for the exam. This book is not intended to replace a textbook in biology. Rather, it is intended to augment textbook and classroom experience in biology and to provide guidance in preparing for this important exam.

Part One of this book is devoted to a brief discussion of the form, structure, and time limitations of the AP examination. It highlights the strategies for success in answering multiple-choice and essay questions in biology and provides a checklist of topics for review. Part Two presents a full-length model exam, patterned on the latest AP examinations in biology, with detailed explanations for all answers and complete instructions for using this exam to evaluate strengths and weaknesses in biology. Part Three, "Important Topics for Review," emphasizes the general concepts of molecular and cellular biology, population biology, and organismic biology that need to be understood in order to earn high scores on the AP test. Part Four, two full-length model examinations, offers the student two additional opportunities to test and evaluate his or her understanding of biology. The Appendix contains useful support materials, including the names of men and women important to biology together with their contributions to the field, a review of the metric system, classic experiments in biology, and a bibliography of suggested books for further study.

PART ONE

ABOUT THE ADVANCED PLACEMENT EXAMINATION

EXAM FACTS

WHAT IS THE ADVANCED PLACEMENT (AP) EXAMINATION?

The Advanced Placement (AP) Examination in Biology is part of a national program that offers you the opportunity to complete college-level studies in biology in your secondary school and to earn college credit upon successful completion of the examination. Many participating high schools offer advanced placement courses in biology. Colleges and universities set their own policies on the use of AP scores to grant appropriate placement or credit for entering freshmen. A list of participating colleges is available from the College Entrance Examination Board; however, specific questions about AP policies are best answered by officials at the college to which you are applying for credit.

Each year over 19,000 students take advantage of the opportunity to earn college credit in biology through the AP program. You can be among them if you work hard and make good use of the opportunity for review and self-evaluation offered by this book.

WHY SHOULD I TAKE THE AP EXAM?

There are three advantages afforded you by the AP program:

1. Saving in Time: You can earn credit for one or two introductory biology courses when you enter college.
2. Saving in Money: Because you may not have to pay for these courses, it is estimated that you could save between $1,000 and $2,000.
3. Increased Probability of Being Accepted to College: By earning an acceptable score on the exam, you will have demonstrated your ability to handle college work.

WHAT IS THE FORMAT OF THE AP EXAM?

The AP Exam is divided into two separately timed tests as shown in the chart below:

TEST	TIME ALLOCATED	NUMBER OF QUESTIONS	DESIGNED TO TEST	PERCENT OF GRADE
Section 1	90 minutes	120 multiple-choice questions	Recall and comprehension of a wide range of ideas and concepts in biology applied to new situations.	60% of total (½% for each multiple-choice question)
Section 2	90 minutes	4 mandatory free-response essay questions	Reasoning and analytical skills and ability to synthesize materials from several sources into a cogent and coherent essay.	40% of total (10% for each essay question)

WHAT IS THE CONTENT OF THE AP EXAM?

The current content and emphasis of the examination is in three areas: Area 1. *Molecules and Cells,* Area 2. *Genetics and Evolution,* Area 3. *Organisms and Populations.*

Section 1 (the multiple-choice section) is constructed to meet the following goals:

Area I. Molecules and Cells (25% of Section 1)

- A. Biological chemistry 7%
- B. Cells 10%
- C. Energy transformations 8%

Area II. Genetics and Evolution (25% of Section 1)

- A. Molecular genetics.................................. 9%
- B. Heredity.................................. 8%
- C. Evolution.................................. 8%

Area III. Organisms and Populations (50% of Section 1)

- A. Principles of Taxonomy.................................. 1%
- B. Survey of Monera, Protista, and Fungi 2%
- C. Plants 15%
- D. Animals.................................. 25%
- E. Ecology 9%

Section 2 (the free-response essay section) is constructed as follows:

Area I. Molecules and Cells (25% of Section 2)
1 essay question

Area II. Genetics and Evolution (25% of Section 2)
1 essay question

Area III. Organisms and Populations (50% of Section 2)
2 essay questions

What Does the AP Exam Test?

Cognition is the mental process by which knowledge is acquired. The AP exam is designed to test the six recognized levels of *cognition:*

1. *Knowledge:* The act of remembering facts, names, dates, principles, etc. *Knowledge* requires nothing more than recall. You are asked to remember some specific fact or general theory.
2. *Comprehension:* The ability to translate materials from one form to another, to explain or summarize materials, and to predict consequences or effects.
3. *Application:* The act of using learned materials in new situations.
4. *Analysis:* The act of breaking down materials into component parts to facilitate understanding of structure, and identification of relationships of parts to one another, or to the whole.
5. *Synthesis:* The ability to put parts together in some unique or creative way.
6. *Evaluation:* The ability to make judgments.

Nongraph multiple-choice questions in Section 1 generally test the first three levels of cognition, i.e., knowledge, comprehension, and application of biological content. Multiple-choice questions which contain graphs (also Section 1) often test analysis, synthesis, and at times, evaluation. Free-response essay questions in Section 2 generally test all levels of cognition with a focus on application, analysis, synthesis, and evaluation.

WHAT KINDS OF QUESTIONS ARE ASKED?

Section 1 (Multiple-choice Questions):

1. Billy is now four years old. Which of the following statements about this child is correct?
   (A) Billy is an unusually large child.
   (B) Billy is an unusually thin child.
   (C) Billy is an unusually small child.
   (D) Billy is an unusually bright child.
   (E) both (A and D)

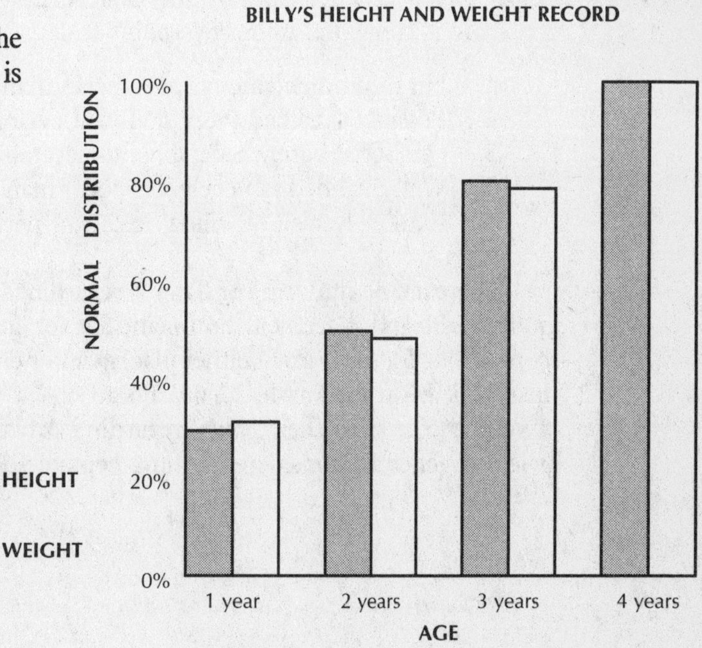

2. Toadstools, mushrooms, puffballs, and bracket fungi belong to the class
   (A) phycomycetes
   (B) ascomycetes
   (C) basidomycetes
   (D) deuteromycetes
   (E) angiomycetes

3. Since water serves as a source of electrons, the lack of water would result in the malfunction of
   (A) photosystem I
   (B) photosystem II
   (C) both (A and B)
   (D) glycolysis
   (E) both (A and D)

Section 2 (Free-response Essay Questions):

1. The three forms of respiration are external, internal, and cellular. Define each form and discuss their interrelationships.
2. The biosphere, that part of earth's surface and atmosphere where living things exist, contains eight recognized biomes. (a) Name four biomes and characterize each; (b) Describe a typical succession; (c) Discuss productivity and stability as they relate to succession.

HOW IS THE AP EXAM SCORED AND HOW ARE THE SCORES USED?

The multiple-choice part of the examination is graded and a score is determined from the total number of correct answers (one point for each correct answer) minus ¼ of a point for each incorrect answer. It is therefore important *not* to guess wildly at answers. The grading of the free-response essay questions involves a complex assessment of facts, definitional terms, examples, concepts, organization, and logic. For a detailed explanation of the process see pages iii and iv of the *Advanced Placement Course Description: Biology*, by the College Entrance Examination Board. The total raw score is converted into a five-point scale.

1 = No recommendations (poor performance).
2 = Possibly qualified (very doubtful performance).
3 = Qualified (barely acceptable to acceptable performance).
4 = Well qualified (a very good performance).
5 = Extremely well qualified (excellent performance).

There are no uniform regulations regarding the use of these scores. Many colleges grant credit and placement automatically for qualified work on the examination (a score of 4 or 5); some grant either placement or credit only; others are still determining their policies in the matter. You should check with the college(s) or university(ies) of your choice as to their policy regarding Advanced Placement scores. A score of 3 or less is generally considered an unacceptable demonstration of proficiency.

TEST BUSTERS

SHORTCUTS AND GUESSING STRATEGIES

GENERAL GUIDELINES

Test Busters will give you suggestions for getting answers to AP questions that you are unable to solve in the usual manner. It will also give you some ways to be more efficient at solving AP problems. Some students may consider "tricking" their way through the test, that is, using multiple-choice testing-taking strategies instead of taking the time to consider each multiple-choice question. This approach is not recommended. Most of the time on the AP exam, the best way to solve a problem is to solve it! Still, there may be times when you will not know how to read a graph, or will not really understand a multiple-choice question, or believe that you do not have enough information to answer an essay question.

Basically, there are three circumstances in which you should resort to guessing strategies.

1. When you are stuck on a problem.
2. When you are short of time.
3. Both of the above.

Guessing Strategies

Guessing strategies are always dependent on the scoring system used to grade the test. The AP exam uses two different scoring systems, one for each section. The score on Section 1, the multiple-choice part of the exam, is determined from the total number of correct answers minus ¼ of a point for each incorrect answer. This scoring system is used to discourage "wild" guessing. In the multiple-choice section of the exam the odds of gaining points are equal to the odds of having points deducted. It does not pay to guess if you are unable to eliminate any of the answers. But the odds of improving your test score are in your favor if you can rule out even one of the answers. The odds in your favor increase as you rule out more answers in any one question.

Free responses to the essay portion of the exam, in Section 2, result in an accumulation of positive score points. These points are based on the number of items that the Educational Testing Service has previously determined should be included in your essay. No points are taken off for incorrect or inappropriate responses. As long as your writing does not take time away from responses that are more likely to be correct, you should respond to every essay question even if you have to guess.

MULTIPLE-CHOICE QUESTIONS

Shortcuts—Using Your Time Effectively

You have 90 minutes to complete Section 1 which contains 120 multiple-choice questions. In precise terms this means that you are to read, interpret, think about, and select one answer for a multiple-choice question every 45 seconds. This may seem impossible but there are several things that you can do to better your chances of using your time effectively.

1. *Practice:* By using the examples in this book, those given out in your preparatory course, class exams, or the purchase of an obsolete exam from the Educational Testing Service, you can become familiar with the format and wording of multiple-choice questions similar to those used in the AP exam. In addition to increasing your information base, taking practice exams can save you time when you need it most.
2. *Words of Totality:* Look out for words like "never" and "always" in multiple-choice questions. It is rarer for an answer to be true in relation to these words of "totality," but it is still a possibility. When you see these words, focus on them and consider them carefully.
3. *Combination Answers:* Some multiple-choice answers will contain statements such as "both A and B" or "all of the above" or "none of these." Do not be distracted by these choices. Multiple-choice questions have only *one* correct answer and do not ask for opinion or personal bias. Quickly go through each choice independently, crossing off the answers that you know are not true. If, after eliminating the incorrect responses, you think there is more than one correct answer, group your answers and see if one of the choices matches yours. If you believe only one answer is correct, do not be distracted by multiple-choice possibilities.

The Fallacy of "First Guess is Best"

Among test-taking folklore there is the principle that your "first guess is best." This piece of folklore is misleading, to say the least. Research indicates that when people change their answers on tests like the AP exam, about two-thirds of the time they go from wrong to right, showing that the first guess is not often the best. Of course, this does not mean that you should feel free to change your answer on a whim; it means that you should be willing to change your answer for a good reason. A test-taker who holds to the "first guess is best" principle might as well not bother to go back to a guessed question, since he or she would be afraid to change it anyway. Remember, your first guess is not better than a result obtained through good, hard, step-by-step, conscious thinking that enables you to select the answer that you believe to be the best.

Three Steps to Follow: Read, Eliminate, Re-read

One of the most helpful strategies for multiple-choice is a three-step process: reading, elimination, and re-reading. Most people tend to see what they expect to see. In test-taking this can be a counterproductive tendency.

1. Read the question quickly but do not skim. It may even pay to quickly read every word. Slow down at words which link by causation such as "due to" and "because" or "as a result of" and at words of totality such as "never" or "always."
2. Eliminate wrong answers one by one. Do not jump to the answer that you think is correct. While elimination may appear to take more time, it is more likely to provide correct answers. In the rush of the test it is easy to select an answer that looks right at first, but on more careful reading does not answer the question. In addition, answer elimination may provide a clue to a misread answer you would have overlooked.
3. Re-read the question, as if you were reading it for the first time. Now choose your answer from your remaining answers based on this re-reading.

SAMPLE MULTIPLE-CHOICE QUESTIONS

Questions 1 through 6 are designed to test the first three levels of cognition—knowledge, comprehension, and application.

1. The structure that is a cell organelle is the
   (A) four-chambered heart
   (B) tubular digestive tract of aves and mammalia
   (C) mitochondrion
   (D) monocot stele
   (E) artery

2. The author of *The Origin of Species* is
   (A) Linus Pauling
   (B) James Watson and Frances Crick
   (C) Gregor Mendel
   (D) Charles Darwin
   (E) Alfred Wallace

3. The cell organelle that would be found in cells requiring large quantities of energy is the
   (A) Golgi apparatus
   (B) mitochondrion
   (C) lysosome
   (D) chromosome
   (E) nucleolus

4. An example of a Phycomycetes is a/an
   (A) angiosperm
   (B) toadstool
   (C) whale
   (D) earthworm
   (E) starfish

5. One cannot drown a grasshopper by holding its head under water because it
   (A) does not need oxygen
   (B) uses its lungs differently from other land animals
   (C) breathes through tracheal tubes that have openings along the abdominal wall
   (D) can hold its breath for several hours
   (E) absorbs oxygen through its gills

6. A metabolic poison that interferes with a particular enzyme of the Krebs citric acid cycle would affect
   (A) the conversion of glucose to pyruvic acid
   (B) the production of lactic acid
   (C) photolysis
   (D) oxidative phosphorylation
   (E) none of these

Question 7 is designed to test the cognitive processes, analysis and synthesis and question 8 tests analysis, synthesis, and evaluation. Questions 7 and 8 are based upon the information provided in the following graph:

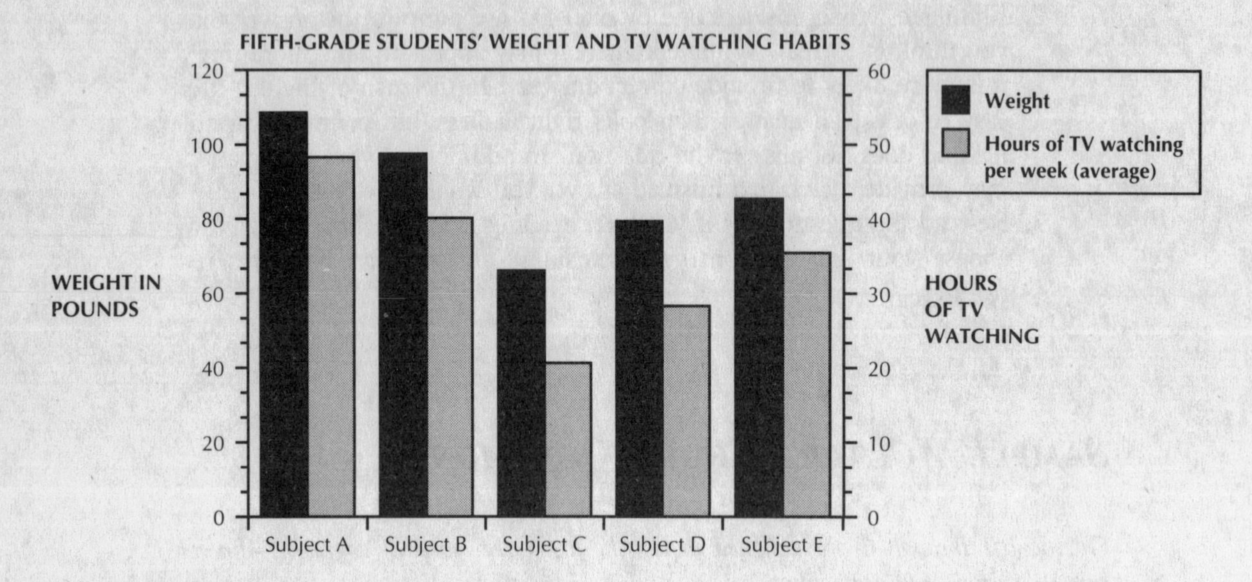

7. Which of the following statements is supported by information provided by the graph?
   (A) The weight of students is inversely proportional to the average number of hours spent watching TV.
   (B) The weight of students is directly proportional to the average number of hours spent watching TV.
   (C) both (A and B)
   (D) No consistent pattern is seen in relation to weight and hours spent watching TV.
   (E) none of these

8. Based on the above graph, one can conclude that
   (A) TV watching decreases the average weight of fifth-grade students
   (B) TV watching increases the average weight of fifth-grade students
   (C) TV watching decreases the weight of some fifth-grade students and increases the weight of others
   (D) TV watching replaces exercise and activity in some fifth-grade students
   (E) none of these

Answers and Explanations for Sample Questions

Questions 1 and 2 test levels of recall only.

1. **(C)** is the only cell organelle on the list. (A) is an animal organ. (B) is part of the digestive system. (D) is the conducting core or xylem and phloem in vascular plants (E) is a blood vessel.

2. **(D)** Charles Darwin wrote *The Origin of Species.*

Questions 3 and 4 test levels of knowledge and comprehension.

3. **(B)** requires you to know what a mitochondrion is and also to understand it to be the cell organelle that functions in energy production. To meet the need for large quantities of energy, a cell must have large numbers of mitochondria.

4. **(B)** is a fungus (nonphotosynthetic plant). The toadstool lacks chlorophyll. (A) is the group of

flowering plants, all of which are photosynthetic. (C), (D), and (E) are all animals.

Questions 5 and 6 test levels of knowledge, comprehension, and the ability to apply the information.

5. **(C)** Gas exchange is important to all living things. In grasshoppers and other insects the gases are exchanged in branched tracheal tubes. (A) is incorrect because all living things require oxygen. (B) and (E) are incorrect since grasshoppers do not have lungs or gills.
6. **(D)** The Krebs citric acid cycle is an integral part of oxidative phosphorylation and therefore any metabolic poison would affect the production of ATP. (A) is glycolysis. (B) is the fermentation pathway that under anaerobic conditions leads to the production of lactic acid. (C) is part of photosynthesis.

Question 7 is designed to test levels of analysis and synthesis.

7. **(B)** is the correct answer. Students of higher weights also exhibit greater numbers of hours watching TV, therefore they are directly proportional. If (A) were correct then students of higher weights would exhibit fewer hours of TV watching. Answer (C) cannot exist, i.e., a number of items cannot have the same variables which vary directly and inversely at the same time. (D) is not true; there is a consistent pattern.

Question 8 is designed to test levels of analysis, synthesis, and evaluation.

8. **(E)** The correct response to this question requires an understanding of the graph combined with the ability to separate correlation from causation, i.e., when one event happens at the same time as another event, it does *not* necessarily cause the other event. It is not possible, therefore, to conclude that TV watching *causes* any response of weight just because the measurement related to one event increases as the other event increases. Therefore, answers (A) through (C) are incorrect. Answer (D) is not based on the information provided by the graph and is also incorrect. The correct answer is (E), none of these.

FREE-RESPONSE ESSAY QUESTIONS

Shortcuts—Using Your Time Effectively

You have 90 minutes to complete Section 2, which contains four free-response essay questions. In precise terms this means that you are to read and interpret, organize data and the supporting information, write an answer in essay format, and proofread an essay (on the average of) every 22.5 minutes. This may seem impossible but there are several things that you can do to better your chances of using your time effectively.

1. *Respond to the Question*
   Key words in the question should guide your responses.

If you are asked to:	Your essay should:
Compare	Show *similarities* between objects, ideas, phenomena, etc.
Contrast	Show *dissimilarities* between two objects, ideas, phenomena, etc.
Define	Provide the accepted definition for a word. The response should be given as a complete sentence.
Describe	Provide a list of features that characterize objects, ideas, phenomena, etc.

Discuss	Select a particular viewpoint and support your position with facts, examples, observations, reasoning, and descriptions.
Explain	Provide a series of well-developed and logical statements which give the reason for or cause of an event or events.
List	Provide a simple series of words, sentences, or phrases as requested. Enhance clarity by labeling each word, sentence, or phrase with sequential numbers or letters.
State	In a logical progression, record the facts related to the question. You are not required to provide proof or illustrations.
Trace	Describe the sequence of the process or the evolutionary development of the event.

2. *Multiple Subquestions:* When you are asked to include several responses in your essay, such as "list, compare, and contrast," respond to each request individually. If you are not asked to integrate your answer, do not do so. Use your time to provide only the information that is requested.
3. *Practice:* In addition to increasing your information base, responding to practice questions can save you time when you need it most.
   By using the examples in this book, those given out in your preparatory course, class exams, textbook review questions, or questions from an obsolete exam available from Educational Testing Service, you can become familiar with the format and wording of essay questions similar to those used in the AP exam. In each practice experience, look for and respond to the key words listed above.

Five Steps to Follow: Read, Quick Response, Re-read, Detailed Response, Re-read

One of the most helpful strategies for handling essay questions is a five-step process which involves: reading, a quick response, re-reading, a detailed response, and re-reading. Most people tend to see what they expect to see. In test-taking, this can be a counterproductive tendency. Many students complete an essay, using up their precious time, only to find that they have answered a question that is different from the one requested on the exam. No matter how well-written the answer, credit is lost when you misinterpret the directional aspects (e.g., key words) of an essay question. In order to minimize that possibility, the following five-step process is recommended.

During the first three to five minutes:

1. *Read* the question quickly but do not skim. It may even pay to quickly read every word. Slow down at key words or other directional aspects of the question. You may want to underline or circle important terms.

2. *Quick Response:* On scrap paper, quickly respond to your first reading of the question in simple terms or short phrases. You may even diagram relationships as they come to you. As you are writing, additional ideas may come to you. Jot those down also. Start to formulate an approach and organization, a logical introduction, body, and closing to the essay.

During the next 15 minutes:

3. *Re-read* the question. Make sure that you are responding to what is requested. If you have underlined or circled terms, do not disregard the unmarked terms.
4. *Detailed Response:* Using the question as your guide, incorporate your quick responses (from your scrap paper) into logical and coherent responses. As you write your detailed response, new ideas may come to you. Interrupt your writing for a moment in order to jot these ideas down. Then return to detailed response. Write in short, declarative sentences. Do not become flowery. If there are multiple subquestions, gauge your time appropriately. Do not spend too much time on one aspect of the response. There are limits to the number of points awarded to any one subresponse. When you are finished with that response, go back to the ideas you jotted down and shape them into responses.

During the last five minutes:

5. *Re-read:* Take time to quickly re-read the question one more time. You may find that you have left out one or more important subquestions. Quickly proofread your response.

Note: When the time alloted for that essay is over, move on to the next essay no matter how much you feel that a few more minutes would improve the essay that you have just completed. If you have time *after* completing the other essays, go back to those which you feel you could improve.

Content of the Essay

Each essay should include an introduction, body, examples, and a conclusion. Work from the broad to the specific. On your scrap paper jot down the following:

1. An introductory sentence which will be the beginning of your first paragraph (introduction). This will state your position or the points that you will make. Later you will fill in the paragraph with supporting and clarifying sentences.
2. Several broad points which will later become the first sentence of each paragraph. This makes up the body of the essay.
3. Examples which prove or illustrate the points mentioned previously. These will be placed in a single paragraph or two or placed where appropriate.
4. A closing sentence. This will be the first sentence of the concluding paragraph and will summarize what you first stated in the introduction and continued to say throughout the essay.

Now begin writing, filling in your paragraphs as you go. *Write in short, declarative sentences.* If you have trouble stating a sentence clearly, try to break it up into smaller sentences. Watch out for your logic, grammar, and spelling.

Fear, Panic, Stage-Fright

The best antidote for fear of test-taking is taking tests. Practice, time yourself on model exams, and as often as possible, face the fear. Deep breathing can sometimes relax some students but the best treatment for test-panic appears to be repeated test-taking.

Practice taking exams in an empty classroom after school, so that the classroom setting is no longer uniquely tied to the final test situation. Try to remember that this is an opportunity to gain time and credits rather than a test of your general worthiness. Concentrate on your test-taking strategies. Repeat your three- and five-step strategies and the essay "key word" definitions to yourself as you prepare to take the exam. It is difficult to think of two things at once. By concentrating on those thoughts which are designed to help you, you should find it more difficult to listen to the thoughts which undermine your confidence.

TOPIC REVIEW CHECKLIST

The Advanced Placement Examination in Biology evaluates your cognitive skills in three broad areas:

- Molecules and Cells (25% of multiple-choice section and 25% of free-response section)
- Genetics and Evolution (25% of multiple-choice section and 25% of free-response section)
- Organisms and Populations (50% of multiple-choice section and 50% of free-response section)

It is important for you to prepare thoroughly in all three areas. Use the checklist provided below to keep track of your study progress. Below is a breakdown of the topics and subtopics which are tested by the AP examination. It provides you with two spaces next to each item on the list. Review each topic. When you feel comfortable with the material, check it off in the first space provided on the checklist. After having made a final review of that topic, check it off in the second space provided.

In addition, each subtopic has a set of specific learning objectives that you should master during your study of that subtopic. It should be noted that this review book is solely intended to serve as a *brief review, a guide for study,* and as *a test* of your ability to succeed on the Advanced Placement Examination in Biology. It is therefore important for you to obtain and use a general college biology textbook as an adjunct to your studies. See the list of suggested textbooks in the Appendix.

MOLECULES AND CELLS

I__:__**Biological chemistry**

A__:__Atoms, bonding, molecules, ions, water, pH

B__:__Carbon, carbohydrates, lipids, proteins, nucleic acids

C__:__Chemical reactions, free-energy changes, dynamic equilibria

D__:__Biological catalysts, cofactors, activity rates, regulation

Major Learning Objectives:

1__:__Describe the structure of the atom and explain the basis of atomic reactivity.

2__:__Define what is meant by acid, salt, and base and explain their roles in biology.
3__:__Discuss the properties of water and its biological importance.
4__:__Name the four major classes of organic compound and briefly discuss the structures and functions of each.
5__:__List a number of examples for each of the four major classes of organic compound.
6__:__Describe each of the four structural arrangements in proteins.
7__:__Discuss the importance of proteins in life.
8__:__Discuss the structure and functions of biological catalysts.
9__:__Discuss the importance of nucleic acids in life.

II__:__**Cells**

A__:__Prokaryotic and eukaryotic cells

B__:__Plant and animal cells

C__:__Cell ultrastructure and functions

D__:__Cell membrane structure and functions

E__:__Cell transports

F__:__Cell cycle: mitosis and cytokinesis

Major Learning Objectives:

1__:__Compare and contrast the typical prokaryotic cell and the typical eukaryotic cell.
2__:__Compare and contrast the typical plant cell and the typical animal cell.
3__:__Using specific names, discuss the historical development of the cell theory.
4__:__Discuss the historical development and current concept of the structure and function of the unit membrane of the cell.
5__:__List and discuss four ways in which materials enter and/or leave the cell.
6__:__Name ten specific cell organelles, discuss the structure of each, state one role for each in the functioning of the cell.
7__:__Describe the cell cycle and characterize each stage of mitosis. Include a statement about the significance of mitosis.

III__:__**Energy transformations in living systems**

A__:__ATP, energy transfer, coupled reactions

B__:__An anabolic process: photosynthesis

C__:__A catabolic process: cellular respiration (aerobic and anaerobic)

Major Learning Objectives:

1__:__Explain how and why ATP serves as the energy molecule for the cell.
2__:__Discuss the historical development of current concepts in photosynthesis.
3__:__Describe the structure of a chloroplast and relate it to photosynthesis.
4__:__List the three major subprocesses of photosynthesis and identify the inputs and outputs of each. Describe the interrelations of each major subprocess.
5__:__List the three major subprocesses of aerobic cellular respiration and identify the inputs and outputs of each. Describe the interrelations of each major subprocess.
6__:__State the sources and numbers of ATP produced from one molecule of glucose.
7__:__Describe and discuss anaerobic respiration (fementation).

IV__:__**Reproduction**

A__:__Mitosis and meiosis

B__:__Fertilization and development in animals and plants

C__:__Genetics

1__:__Mendelian genetics
2__:__Modern genetics
3__:__Molecular genetics

Major Learning Objectives:

1__:__Characterize each stage of the cell cycle and discuss the importance and significance of the cycle.
2__:__Compare and contrast mitosis and meiosis.
3__:__Compare and contrast oogenesis and spermatogenesis.
4__:__Discuss the forms of asexual and sexual reproduction.
5__:__Compare and contrast external and internal fertilization.
6__:__Discuss the contributions of Gregor Mendel to the science of genetics.
7__:__Discuss the importance of Drosophila and bacteria in the development of modern concepts in genetics.
8__:__List and discuss non-Mendelian genetic concepts.

GENETICS AND EVOLUTION

I__:__**Molecular genetics**

A__:__DNA structure, function, and replication

B__:__Chromosomal structure, nucleosomes, transportable elements

C__:__Transcription, translation

D__:__Regulation of gene expression, mutations

E__:__Genetic engineering: recombinant DNA, DNA cloning and hybridization, DNA sequencing

F__:__DNA and RNA viruses

Major Learning Objectives:

1__:__Contrast and compare DNA and RNA structure.
2__:__Explain how DNA replicates and how it relates to heredity.
3__:__Compare and contrast transcription and translation.
4__:__Describe the process of protein synthesis.
5__:__Describe two prokaryotic models for the regulation of gene expression.
6__:__Name four types of chromosome mutation and give examples of each.
7__:__For each of the following, describe the process involved and state the purpose of: Recombinant DNA techniques, DNA cloning techniques, DNA hybridization techniques, DNA sequencing techniques.
8__:__Compare and contrast: DNA and RNA viruses, viral lytic, and lysogenic cycles.

II__:__**Heredity**

A__:__Cell cycle: Meiosis

B__:__Mendelian genetics

C__:__Modern genetics: chromosomes, genes, alleles, human genetics

Major Learning Objectives:

1__:__Describe the process of meiosis.
2__:__Compare and contrast mitosis and meiosis.
3__:__Compare and contrast human oogenesis and spermatogenesis.
4__:__State the principles of Mendelian genetics.
5__:__Discuss the importance of Drosophila and bacteria in the development of modern genetics.
6__:__State and discuss non-Mendelian genetic concepts.
7__:__Name and describe five human genetic diseases that are the result of dominant genes.
8__:__Name and describe five human genetic diseases that are the result of recessive genes.

III. Evolution

A. Origin of life

B. Evidence for evolution

C. Darwin versus Lamarck, natural selection

D. Hardy–Weinberg principle, allelic frequencies

E. Origin of species: isolation, allopatry, sympatry, adaptive radiation

F. Evolution patterns

Major Learning Objectives:

1. Describe and explain the current concept of the origin of life on earth.
2. Describe the evidences in support of the current view of organic evolution.
3. Define natural selection, explain its role in evolution, and give three modern examples of natural selection.
4. State the Hardy–Weinberg Law and state its importance to evolution.
5. Discuss the mechanisms that allow for speciation.
6. Define adaptive radiation and discuss its role in evolution.
7. Discuss primate evolution.

ORGANISMS AND POPULATIONS

I. Taxonomic principles and systematics

A. Five kingdom system

B. Human classification

Major Learning Objectives:

1. Name the five kingdoms and state the characteristic(s) of each.
2. In descending order, list the major taxa used in classifying plants and animals.
3. Discuss the classification of humans

II. Monera, Protista, and Fungi

Major Learning Objectives:

1. For each of the biological kingdoms above, state its characteristic(s) and give examples of each.

III__:__Plants

A__:__Classification, phylogeny, adaptation to various environments, alternations of generations

B__:__Structure and function of vascular plants

C__:__Plant reproduction: seed formation, germination, growth

D__:__Plant growth regulation: hormones

E__:__Plant responses: tropisms, photoperiodicity

Major Learning Objectives:

1__:__Name the major phyla in the plant kingdom and give an example of each. For each phylum, name the classes and give examples.
2__:__For each, describe the adaptations of xerophyte, hydrophytes, and mesophytes.
3__:__Discuss the pattern of alternations of generations seen in the evolution of plants.
4__:__Compare and contrast monocots and dicots both externally and internally.
5__:__Discuss vascular transport in plants.
6__:__Discuss meristem types and functions in plants.
7__:__Describe and discuss the reproductive process in vascular plants.
8__:__Name four classes of plant hormone and state the function(s) of each.
9__:__Explain periodic flowering in plants.

IV__:__Animals

A__:__Classification, phylogeny, major phyla

B__:__Tissue, organ, and system structure and function

C__:__Vertebrate homeostasis, immune response

D__:__Reproduction: gametogenesis, fertilization, development

E__:__Behavior: innate and learned behavior, biological clocks, social behavior

Major Learning Objectives:

1__:__Name the major phyla in the animal kingdom and state the characteristics of each.
2__:__For each major animal phylum, list the major classes and state the characteristics of each.
3__:__Compare and contrast monotremes, marsupials, and placental mammals.

4__:__List and characterize the primary tissues of the higher animal body.

5__:__Skeletal System

- a__:__Describe the structure of the vertebrate skeleton.
- b__:__Discuss the role of hormones in calcium deposition in bone.
- c__:__Name two types of support tissue in higher animals.

6__:__Muscle System

- a__:__Name three types of muscle and state the function of each.
- b__:__Describe the structure of the sarcomere and explain its role in contraction of skeletal muscle.
- c__:__Discuss the role of antagonism in the muscle system.

7__:__Digestive System

- a__:__List three forms of animal nutrition and describe each.
- b__:__Discuss the evolution of the digestive system and provide examples.
- c__:__List and discuss five basic functions of the human digestive system.
- d__:__In sequence, list the structures of the human digestive system and discuss the digestive role of each.
- e__:__Discuss the roles of the liver and the pancreas in digestion.

8__:__Circulatory System

- a__:__List seven functions of the circulatory system.
- b__:__Discuss the evolution of the circulatory system and provide examples.
- c__:__Compare and contrast open and closed circulation.
- d__:__Discuss two evolutionary trends observed in the vertebrate circulatory system.
- e__:__Discuss the evolution of the vertebrate heart.
- f__:__Discuss the composition of the higher vertebrate blood.
- g__:__List the cell types found in human blood and state the function(s) of each.
- h__:__Contrast blood clotting and blood agglutination.
- i__:__Discuss the role of the lymphatic system in higher vertebrates.
- j__:__Discuss compatibility as it relates to ABO and Rh blood groups in humans.
- k__:__Describe the flow of blood through the heart and list all structures involved.

9__:__Excretory System

a__:__Discuss the evolution of the excretory system. Provide examples.

b__:__Describe the structure of the human kidney.

c__:__Describe the structure of the human excretory system.

d__:__Discuss the role of hormones in the control of human excretion.

10__:__Nervous System

a__:__List three types of structural neuron and three types of functional neuron.

b__:__Discuss the evolution of the nervous system.

c__:__Describe the composition of nervous tissue.

d__:__Explain how an impulse is generated, transmitted, and moves across a synaptic junction.

e__:__Describe the structure and function of the parts of the human nervous system.

f__:__List and describe five types of general sense receptors in humans.

g__:__Describe the structure of the human ear and discuss the mechanism of hearing.

h__:__Describe the structure of the human eye and discuss the mechanism of vision.

11__:__Endocrine System

a__:__List the endocrine glands of the human body, name the hormones produced by each, and state the general function(s) of each hormone.

b__:__Describe the proposed mechanism(s) for hormone action.

12__:__Respiratory System

a__:__Compare and contrast the three forms of respiration.

b__:__List the four major structures used for respiration in animals.

c__:__Describe the structure of the human respiratory system and explain how it functions.

d__:__Explain the mechanism of breathing in the human lung.

13__:__Reproductive System

a__:__Discuss the forms of reproduction found in animals.

b__:__Describe both human male and female reproductive structures and explain how they function.

c__:__Discuss the role of hormones in the female reproductive system.

d__:__Discuss the role of hormones in the male reproductive system.

e__:__Discuss gametogenesis in humans.

14__:__Behavioral Biology

a__:__List five types of learning and explain each.
b__:__Discuss the role of chemical communication in behavior.
c__:__List seven behavioral states and explain each.

V__:__**Ecology**

A__:__Populations: dynamics, biotic potential, limiting factors

B__:__Ecosystem energy flow and productivity

C__:__Communities and ecosystems interactions

D__:__Ecocycles: nitrogen, carbon, and phosphorous

Major Learning Objectives:

1__:__Discuss population dynamics in an ecosystem.
2__:__Describe: an aquatic food chain, a terrestrial food chain. Name the producer, the various levels of consumer, and the decomposer. Discuss the chemical cycles involved.
3__:__Describe energy flow through an ecosystem.
4__:__Describe a food web and explain how pyramids describe the size of various trophic level in that web.
5__:__Describe the carbon cycle, the nitrogen cycle, the water cycle, and the phosphorous cycle and explain the human influences on each cycle.
6__:__List three principal categories of community interaction and provide an example of each.
7__:__List three types of symbiosis and provide one example of each.
8__:__Explain what is meant by ecological succession and provide one example. Name eight major biomes and describe each.

PART TWO

DIAGNOSTIC MODEL EXAMINATION

The examination that follows offers a chance for you to assess your readiness for the Advanced Placement Examination (AP) in Biology. Allow yourself 90 minutes to answer all of the 120 multiple-choice questions in Section 1 and 90 minutes to answer the four essays in questions in Section 2. When you have completed the exam, check your answers against the Answer Key and Explanatory Answers at the end of the exam. Complete the Diagnostic Chart on page 48 to identify any remaining weaknesses.

Use the specially constructed answer sheet to record your answers for Section 1. Use plain or lined paper to answer the free-response questions in Section 2. Sample answers are provided for each essay question.

MODEL EXAMINATION, SECTION 1
ANSWER SHEET

1 Ⓐ Ⓑ Ⓒ Ⓓ Ⓔ
2 Ⓐ Ⓑ Ⓒ Ⓓ Ⓔ
3 Ⓐ Ⓑ Ⓒ Ⓓ Ⓔ
4 Ⓐ Ⓑ Ⓒ Ⓓ Ⓔ
5 Ⓐ Ⓑ Ⓒ Ⓓ Ⓔ
6 Ⓐ Ⓑ Ⓒ Ⓓ Ⓔ
7 Ⓐ Ⓑ Ⓒ Ⓓ Ⓔ
8 Ⓐ Ⓑ Ⓒ Ⓓ Ⓔ
9 Ⓐ Ⓑ Ⓒ Ⓓ Ⓔ
10 Ⓐ Ⓑ Ⓒ Ⓓ Ⓔ
11 Ⓐ Ⓑ Ⓒ Ⓓ Ⓔ
12 Ⓐ Ⓑ Ⓒ Ⓓ Ⓔ
13 Ⓐ Ⓑ Ⓒ Ⓓ Ⓔ
14 Ⓐ Ⓑ Ⓒ Ⓓ Ⓔ
15 Ⓐ Ⓑ Ⓒ Ⓓ Ⓔ
16 Ⓐ Ⓑ Ⓒ Ⓓ Ⓔ
17 Ⓐ Ⓑ Ⓒ Ⓓ Ⓔ
18 Ⓐ Ⓑ Ⓒ Ⓓ Ⓔ
19 Ⓐ Ⓑ Ⓒ Ⓓ Ⓔ
20 Ⓐ Ⓑ Ⓒ Ⓓ Ⓔ
21 Ⓐ Ⓑ Ⓒ Ⓓ Ⓔ
22 Ⓐ Ⓑ Ⓒ Ⓓ Ⓔ
23 Ⓐ Ⓑ Ⓒ Ⓓ Ⓔ
24 Ⓐ Ⓑ Ⓒ Ⓓ Ⓔ
25 Ⓐ Ⓑ Ⓒ Ⓓ Ⓔ
26 Ⓐ Ⓑ Ⓒ Ⓓ Ⓔ
27 Ⓐ Ⓑ Ⓒ Ⓓ Ⓔ
28 Ⓐ Ⓑ Ⓒ Ⓓ Ⓔ
29 Ⓐ Ⓑ Ⓒ Ⓓ Ⓔ
30 Ⓐ Ⓑ Ⓒ Ⓓ Ⓔ
31 Ⓐ Ⓑ Ⓒ Ⓓ Ⓔ
32 Ⓐ Ⓑ Ⓒ Ⓓ Ⓔ
33 Ⓐ Ⓑ Ⓒ Ⓓ Ⓔ
34 Ⓐ Ⓑ Ⓒ Ⓓ Ⓔ
35 Ⓐ Ⓑ Ⓒ Ⓓ Ⓔ
36 Ⓐ Ⓑ Ⓒ Ⓓ Ⓔ
37 Ⓐ Ⓑ Ⓒ Ⓓ Ⓔ
38 Ⓐ Ⓑ Ⓒ Ⓓ Ⓔ
39 Ⓐ Ⓑ Ⓒ Ⓓ Ⓔ
40 Ⓐ Ⓑ Ⓒ Ⓓ Ⓔ
41 Ⓐ Ⓑ Ⓒ Ⓓ Ⓔ
42 Ⓐ Ⓑ Ⓒ Ⓓ Ⓔ
43 Ⓐ Ⓑ Ⓒ Ⓓ Ⓔ
44 Ⓐ Ⓑ Ⓒ Ⓓ Ⓔ
45 Ⓐ Ⓑ Ⓒ Ⓓ Ⓔ
46 Ⓐ Ⓑ Ⓒ Ⓓ Ⓔ
47 Ⓐ Ⓑ Ⓒ Ⓓ Ⓔ
48 Ⓐ Ⓑ Ⓒ Ⓓ Ⓔ
49 Ⓐ Ⓑ Ⓒ Ⓓ Ⓔ
50 Ⓐ Ⓑ Ⓒ Ⓓ Ⓔ
51 Ⓐ Ⓑ Ⓒ Ⓓ Ⓔ
52 Ⓐ Ⓑ Ⓒ Ⓓ Ⓔ
53 Ⓐ Ⓑ Ⓒ Ⓓ Ⓔ
54 Ⓐ Ⓑ Ⓒ Ⓓ Ⓔ
55 Ⓐ Ⓑ Ⓒ Ⓓ Ⓔ
56 Ⓐ Ⓑ Ⓒ Ⓓ Ⓔ
57 Ⓐ Ⓑ Ⓒ Ⓓ Ⓔ
58 Ⓐ Ⓑ Ⓒ Ⓓ Ⓔ
59 Ⓐ Ⓑ Ⓒ Ⓓ Ⓔ
60 Ⓐ Ⓑ Ⓒ Ⓓ Ⓔ
61 Ⓐ Ⓑ Ⓒ Ⓓ Ⓔ
62 Ⓐ Ⓑ Ⓒ Ⓓ Ⓔ
63 Ⓐ Ⓑ Ⓒ Ⓓ Ⓔ
64 Ⓐ Ⓑ Ⓒ Ⓓ Ⓔ
65 Ⓐ Ⓑ Ⓒ Ⓓ Ⓔ
66 Ⓐ Ⓑ Ⓒ Ⓓ Ⓔ
67 Ⓐ Ⓑ Ⓒ Ⓓ Ⓔ
68 Ⓐ Ⓑ Ⓒ Ⓓ Ⓔ
69 Ⓐ Ⓑ Ⓒ Ⓓ Ⓔ
70 Ⓐ Ⓑ Ⓒ Ⓓ Ⓔ
71 Ⓐ Ⓑ Ⓒ Ⓓ Ⓔ
72 Ⓐ Ⓑ Ⓒ Ⓓ Ⓔ
73 Ⓐ Ⓑ Ⓒ Ⓓ Ⓔ
74 Ⓐ Ⓑ Ⓒ Ⓓ Ⓔ
75 Ⓐ Ⓑ Ⓒ Ⓓ Ⓔ
76 Ⓐ Ⓑ Ⓒ Ⓓ Ⓔ
77 Ⓐ Ⓑ Ⓒ Ⓓ Ⓔ
78 Ⓐ Ⓑ Ⓒ Ⓓ Ⓔ
79 Ⓐ Ⓑ Ⓒ Ⓓ Ⓔ
80 Ⓐ Ⓑ Ⓒ Ⓓ Ⓔ
81 Ⓐ Ⓑ Ⓒ Ⓓ Ⓔ
82 Ⓐ Ⓑ Ⓒ Ⓓ Ⓔ
83 Ⓐ Ⓑ Ⓒ Ⓓ Ⓔ
84 Ⓐ Ⓑ Ⓒ Ⓓ Ⓔ
85 Ⓐ Ⓑ Ⓒ Ⓓ Ⓔ
86 Ⓐ Ⓑ Ⓒ Ⓓ Ⓔ
87 Ⓐ Ⓑ Ⓒ Ⓓ Ⓔ
88 Ⓐ Ⓑ Ⓒ Ⓓ Ⓔ
89 Ⓐ Ⓑ Ⓒ Ⓓ Ⓔ
90 Ⓐ Ⓑ Ⓒ Ⓓ Ⓔ
91 Ⓐ Ⓑ Ⓒ Ⓓ Ⓔ
92 Ⓐ Ⓑ Ⓒ Ⓓ Ⓔ
93 Ⓐ Ⓑ Ⓒ Ⓓ Ⓔ
94 Ⓐ Ⓑ Ⓒ Ⓓ Ⓔ
95 Ⓐ Ⓑ Ⓒ Ⓓ Ⓔ
96 Ⓐ Ⓑ Ⓒ Ⓓ Ⓔ
97 Ⓐ Ⓑ Ⓒ Ⓓ Ⓔ
98 Ⓐ Ⓑ Ⓒ Ⓓ Ⓔ
99 Ⓐ Ⓑ Ⓒ Ⓓ Ⓔ
100 Ⓐ Ⓑ Ⓒ Ⓓ Ⓔ
101 Ⓐ Ⓑ Ⓒ Ⓓ Ⓔ
102 Ⓐ Ⓑ Ⓒ Ⓓ Ⓔ
103 Ⓐ Ⓑ Ⓒ Ⓓ Ⓔ
104 Ⓐ Ⓑ Ⓒ Ⓓ Ⓔ
105 Ⓐ Ⓑ Ⓒ Ⓓ Ⓔ
106 Ⓐ Ⓑ Ⓒ Ⓓ Ⓔ
107 Ⓐ Ⓑ Ⓒ Ⓓ Ⓔ
108 Ⓐ Ⓑ Ⓒ Ⓓ Ⓔ
109 Ⓐ Ⓑ Ⓒ Ⓓ Ⓔ
110 Ⓐ Ⓑ Ⓒ Ⓓ Ⓔ
111 Ⓐ Ⓑ Ⓒ Ⓓ Ⓔ
112 Ⓐ Ⓑ Ⓒ Ⓓ Ⓔ
113 Ⓐ Ⓑ Ⓒ Ⓓ Ⓔ
114 Ⓐ Ⓑ Ⓒ Ⓓ Ⓔ
115 Ⓐ Ⓑ Ⓒ Ⓓ Ⓔ
116 Ⓐ Ⓑ Ⓒ Ⓓ Ⓔ
117 Ⓐ Ⓑ Ⓒ Ⓓ Ⓔ
118 Ⓐ Ⓑ Ⓒ Ⓓ Ⓔ
119 Ⓐ Ⓑ Ⓒ Ⓓ Ⓔ
120 Ⓐ Ⓑ Ⓒ Ⓓ Ⓔ

MODEL EXAMINATION 1

SECTION 1

Time—90 minutes

Number of Questions—120

Percent of Total Grade—60

Directions: **For each of the 120 questions or incomplete statements below, select the choice from (A–E) that best answers the question or completes the statement. Record your answers on the answer sheet provided.**

1. Select the statement that is correct.
   (A) Photosynthesis is a catabolic activity that breaks down glucose.
   (B) The Calvin Cycle requires molecular carbon dioxide, ATP, and molecular oxygen in order to manufacture glucose.
   (C) Glycolysis is aerobic during fermentation.
   (D) Most of the ATP in cellular respiration is produced anaerobically.
   (E) The Calvin Cycle is a light-independent metabolism.

2. The green outer ring of leaves in the flower structure is composed of
   (A) petals
   (B) sepals
   (C) stamens
   (D) carpels
   (E) stigmas

3. The best evidence currently supports the hypothesis that DNA replication occurs by means of
   (A) photographic replication
   (B) semiconservative replication
   (C) protein synthesis
   (D) conservative replication
   (E) dispersive replication

Questions 4 and 5. In 1965 two male and two female deer were brought to Block Island, Rhode Island. These four deer were a seed population that encountered no serious predators. Since that time the deer population on the island has followed the pattern depicted by the graph on page 30.

4. The best explanation of the graph from 1967 to 1975 is:
   (A) The deer population was in a growth phase of a typical growth curve.
   (B) The deer during this period were afraid of the people and cars and chose to stay in the bushes rather than venture into the gardens.
   (C) There was a severe drought during the period.
   (D) The deer population was stable.
   (E) The deer population was on a decline because of the decrease in food.

5. According to the graph, the period during which deer hunting was probably legalized is
   (A) 1969–1972
   (B) 1975–1977
   (C) 1977–1980
   (D) 1983–1985
   (E) 1988–1991

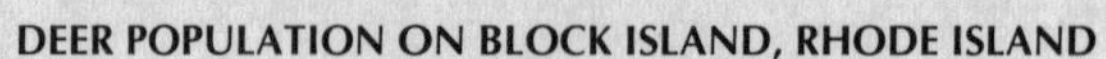

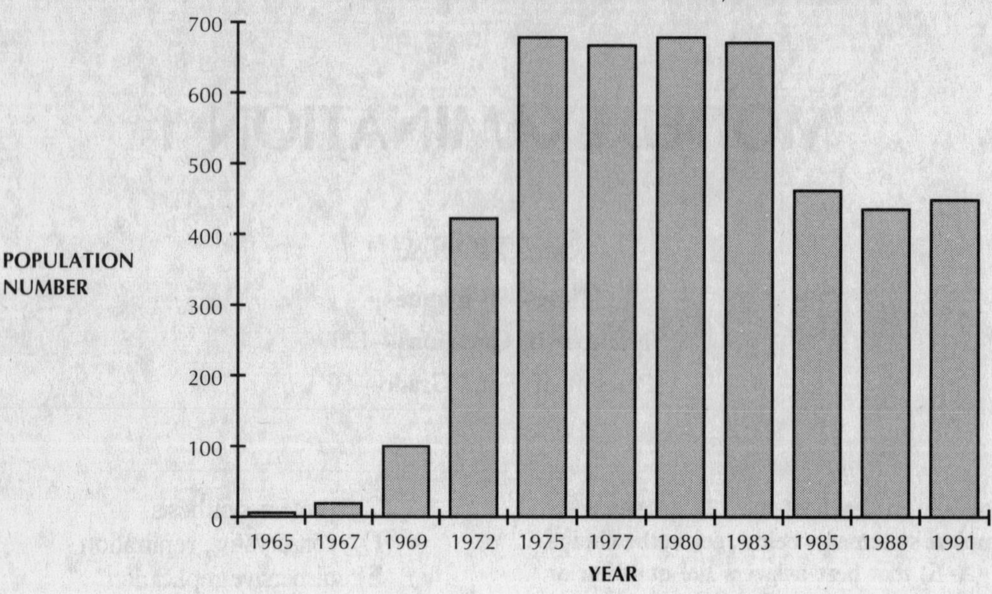

6. All enzymes are known to be
   (A) tertiary proteins with specific active sites
   (B) polysaccharides in plant cells but carbohydrates in animal cells
   (C) capable of self-reproduction exclusive of any other living system
   (D) long chains of nucleotides
   (E) none of these

7. The concept that all life on earth arose by spontaneous generation from simple inorganic molecules in the environment was first proposed by
   (A) James Watson and Francis Crick
   (B) Charles Darwin
   (C) Gregor Mendel
   (D) Alexander Oparin
   (E) Robert Hooke

8. Primary producers on earth are
   (A) hererotrophs (D) carnivores
   (B) autotrophs (E) herbivores
   (C) parasites

Questions 9 and 10 are based on the chart on page 31. In humans, hemoglobin is normally 70% oxygen saturated at a partial pressure of 40 mm Hg. Increasing the acidity reduces the percentage of oxygen saturation.

9. The curve representing the normal percentage of oxygen saturation of hemoglobin is
   (A) curve A (C) curve C (E) curve E
   (B) curve B (D) curve D

10. The curve representing the greatest acidity is
   (A) curve A (C) curve C (E) curve E
   (B) curve B (D) curve D

11. Endogenous cycles that are generally not dependent upon environmental factors are
   (A) reflexes
   (B) circadian rhythms
   (C) seasonal changes
   (D) heartbeat and blood pressure
   (E) sleep and wakefulness

12. Experimental evidence in support of the concept that life arose by spontaneous generation from simple, inorganic molecules was first provided by
   (A) Alexander Oparin
   (B) Joseph Haldane
   (C) Charles Darwin
   (D) Sidney Fox
   (E) Stanley Miller

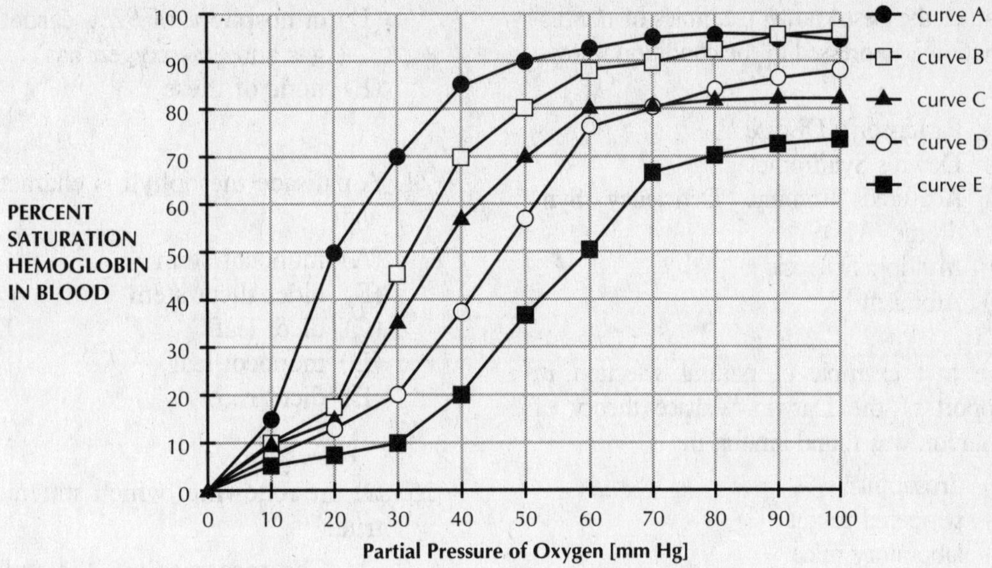
PERCENT
SATURATION
HEMOGLOBIN
IN BLOOD
100
90
80
70
60
50
40
30
20
10
0
0
10
20
30
40
50
60
70
80
90
100
Partial Pressure of Oxygen [mm Hg]
curve A
curve B
curve C
curve D
curve E

13. The most reasonable explanation to account for migration and homing in some animals is
    (A) animals guide themselves by the stars
    (B) animals can guide themselves by the magnetic lines created by the earth itself
    (C) the animals pick out certain landmarks along the way
    (D) one animal follows another along the way
    (E) the animals are all born with genetic maps of the locations of the migratory paths

14. The Second Law of Thermodynamics predicts that
    (A) energy may be changed from one form into another
    (B) there is a reduction in the amount of free energy every time energy is changed from one form into another
    (C) energy can neither be created nor destroyed
    (D) energy in a black hole is being created
    (E) endergonic reactions will produce more energy than they consume

15. The fixation of carbon dioxide into energy-rich glucose occurs during
    (A) the Krebs Citric Acid Cycle
    (B) glycolysis
    (C) noncyclic photophosphorylation
    (D) the Calvin Cycle
    (E) the electron transport system production of ATP

16. One of the most important animal polysaccharides is
    (A) glucose
    (B) galactose
    (C) ribulose
    (D) glycogen
    (E) starch

17. The element that serves as the structural basis of all organic compounds is
    (A) hydrogen
    (B) water
    (C) carbon
    (D) nitrogen
    (E) oxygen

18. Enzymes that lack a prosthetic group require
    (A) activation energy in order to function
    (B) a pH of less than 3.0 in order to function
    (C) a metallic ion or some nonproteinaceous organic substance in order to function
    (D) substrates common to all enzymes
    (E) none of these

19. One of the best-known examples of nondisjunction is expressed in the condition known as
    (A) Parkinson's Disease
    (B) Down's Syndrome
    (C) Acquired Immune Deficiency Syndrome (AIDS)
    (D) Multiple Sclerosis
    (E) Albinism

20. The first example of natural selection in support of the Darwin–Wallace theory of evolution was found among the
    (A) drosophila
    (B) peppered moths
    (C) laboratory mice
    (D) wheat coleoptiles
    (E) bacterial mutants

21. One hypothesis used to explain how the first cells on earth were formed is
    (A) coacervate droplet condensation resulting in the spontaneous generation of cells
    (B) sexual reproduction between common nucleic acids
    (C) photosynthetic nodules coalesing with mitochondria
    (D) autotrophic precursors developing from green pigment natural in southern waters
    (E) none of these

22. The industrial melanism exhibited by the peppered moth in England is an example of
    (A) parallel evolution
    (B) evolutionary adaptation
    (C) divergent evolution
    (D) convergent evolution
    (E) codominant genes

23. The primitive atmosphere of earth prior to the development of life is thought to have been a/an
    (A) oxidizing atmosphere
    (B) reducing atmosphere
    (C) atmosphere similar to today's atmosphere
    (D) atmosphere of 92% carbon dioxide gas and 8% oxygen gas
    (E) none of these

24. A palisade mesophyll is characteristic of a/an
    (A) monocot stem
    (B) older dicot stem
    (C) dicot leaf
    (D) monocot leaf
    (E) dicot root

25. Of the following, which statement is *not* true?
    (A) Chromosomes are the units of heredity.
    (B) Centrioles are composed of microtubules.
    (C) Lysosomes contain moderate quantities of nucleic acids.
    (D) The golgi apparatus is particularly well-developed in endocrine gland cells.
    (E) Chloroplasts are not found in fungi cells.

26. Photophosphorylation and photolysis are both photosynthetic events that occur in
    (A) chloroplasts
    (B) chromosomes
    (C) mitochondria
    (D) ribosomes
    (E) peroxisomes

27. Telophase in a plant cell can be identified as a time of
    (A) chromosomal doubling and coiling
    (B) cell plate formation perpendicular to the axis of the spindle fibers
    (C) crossing over among the chromosomes
    (D) pinching of the cytoplasmic membrane
    (E) DNA replication

28. The primary structure of a protein is represented by the
    (A) sequence of amino acids in the molecule
    (B) pleating or coiling in the molecule
    (C) spatial folding of the molecule
    (D) aggregation of several tertiary proteins
    (E) sequence of deoxyribonucleic acid nucleotides

Questions 29 and 30. The graph below deals with a comparison of human populations in developed and developing countries.

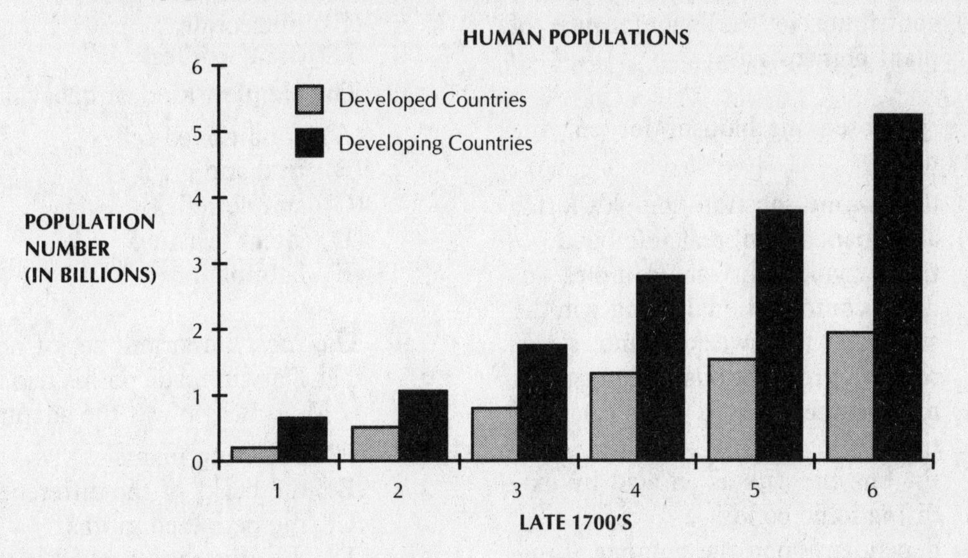

29. If you were a biology teacher, you would use this graph in a class in:
    (A) behavior
    (B) digestion
    (C) ecology
    (D) cellular respiration
    (E) plant anatomy

30. The conclusion that can be drawn from this graph is that
    (A) populations in developing countries are growing faster than they are growing in developed countries
    (B) populations with increased development decrease in size
    (C) populations in developed countries are growing faster than they are growing in developing countries
    (D) populations around the world grow equally
    (E) development appears to have no effect on the growth of populations

31. All enzymes must possess
    (A) nucleotides
    (B) amino acids
    (C) fatty acids and alcohol
    (D) monosaccharides and disaccharides
    (E) dicarboxylic acid and dopamine

32. Allosteric inhibition may involve
    (A) an enzyme that exists in one of two interchangeable conformational forms
    (B) stabilization of an enzyme so that it is incapable of accepting a substrate
    (C) a cofactor that must couple with a carboxylic and amine group
    (D) a reduction in the number of amine groups in a protein substrate
    (E) none of these

33. Molecular oxygen resulting from photolysis of water is a byproduct of
    (A) glycolysis
    (B) the electron transport system

(C) cyclic photophosphorylation
(D) noncyclic photophosphorylation
(E) cellular respiration

34. Plant hormones generally act to
   (A) stimulate cell elongation
   (B) inhibit cell division
   (C) promote flowering
   (D) inhibit RNA and protein synthesis
   (E) contribute to the maintenance of plant homeostasis

35. One proposed mechanism for enzyme action is
   (A) the enzyme–substrate complex forms a permanent, unbreakable bond
   (B) the enzyme–substrate complex induces conformational changes in the substrate that weaken and strain reactive group bonds making them more susceptible to chemical reaction
   (C) the enzyme acts as an acid by oxidizing ionic bonds
   (D) dependent upon the number of mitochondria per cell
   (E) the formation of electrostatic salt linkages

36. Darwin's Theory of Evolution by Natural Selection is *not* based upon which of the following observations?
   (A) There is variation within populations of organisms.
   (B) Many differences between and among the members of a species are inherited.
   (C) Females produce many more offspring than live to mature sexually and reproduce.
   (D) Acquired characteristics are inherited.
   (E) none of these

37. The first cells to have evolved were probably
   (A) procaryotic heterotrophs
   (B) autotrophs
   (C) photosynthetic animals and plants
   (D) eucaryotes
   (E) sexually reproducing

38. A multinucleate animal cell is called a
   (A) histone
   (B) coenocyte
   (C) syncytium
   (D) centromere
   (E) nucleotide

39. The simplest kind of behavior is
   (A) conditioned reflex
   (B) trial and error
   (C) simple reflex
   (D) latent learning
   (E) habituation

40. The most important set of observations that Darwin made on his trip aboard the H.M.S. Beagle was the adaptations of
   (A) flowering plants
   (B) the beaks of the different finches
   (C) the peppered moths
   (D) the crustaceans on all of the beaches
   (E) the necks of the various giraffes that he saw

Questions 41 and 42 are based on the graph on page 35. Nitrogen dioxide is an oxidant gas that is capable of producing emphysema in hamsters exposed to the gas at various concentrations for extended periods of time.

41. Which of the following statements is correct?
   (A) Mild exposure to nitrogen dioxide appears to have no effect on the survival of hamsters.
   (B) The group that is most likely to survive continual exposure to nitrogen dioxide is the group exposed to 7 ppm.
   (C) Week four appears to be the time when nitrogen dioxide exposure has its greatest effect.

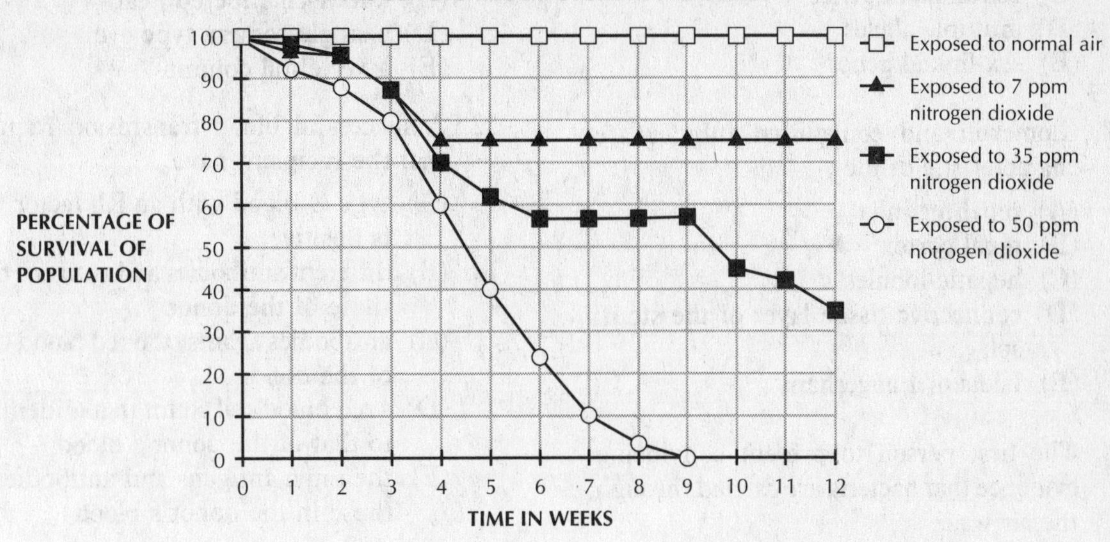

(D) The concentration of nitrogen dioxide plays no role in the percentage of survival of hamsters exposed to nitrogen dioxide.
(E) After six weeks of exposure to nitrogen dioxide more hamsters survive at a 35 ppm exposure than do those exposed to 7 ppm.

42. According to the data in the graph
(A) the effect of nitrogen dioxide gas on hamsters appears to be directly proportional to the concentration of the gas
(B) emphysema and lung fibrosis are not produced by molecular oxygen, ozone, or bleomycin
(C) the effect of nitrogen dioxide gas on hamsters appears to be inversely proportional to the concentration of the gas
(D) the heart rate of the hamsters is greatest when they are exposed to concentrations of 50 ppm nitrogen dioxide
(E) at week six more than 75% of the hamsters exposed to 7 ppm nitrogen dioxide have died

43. Given the element nitrogen, which has an atomic number of 14 and an atomic weight of 29, one can conclude that nitrogen
(A) contains 14 neutrons and 15 protons
(B) is a nonreactive compound as it cannot share electrons
(C) contains 15 neutrons and 14 protons
(D) will readily combine with neon
(E) has 43 neutrons

44. A small agar block that has been impregnated with the plant hormone auxin (IAA) applied to the side of a coleoptile tip will cause
(A) a bud to develop in the region of the application
(B) the root of the plant to bend up towards the light
(C) the coleoptile to bend towards the side where the agar block is attached
(D) the coleoptile to bend away from the side where the agar block is attached
(E) a lateral stem to develop below the region of the application

45. Tetrad formation and crossing-over during meiosis occurs at the time of
(A) telophase I
(B) telophase II
(C) metaphase I
(D) metaphase II
(E) growth I of interphase

46. Metabolic diseases are often the result of
(A) dominant gene expression
(B) recessive gene expression

(C) codominant genes
(D) multiple alleles
(E) sex-linked genes

47. Glomeruli and convoluted tubules are characteristic of the

(A) renal medulla
(B) renal cortex
(C) hepatic lobules
(D) connective tissue layer of the stomach
(E) Islets of Langerhans

48. The first person to present convincing evidence that bacteria are carried through the air was

(A) Rudolph Virchow
(B) William Harvey
(C) James Watson
(D) Louis Pasteur
(E) Charles Darwin

49. The three principal steps in the nitrogen cycle are

(A) ammonification, nitrification, and ecological succession
(B) ammonification, assimilation, and phosphorylation
(C) ammonification, nitrification, and assimilation
(D) phosphorylation, hydration, and assimilation
(E) respiration, decarboxylation, and quantification

50. Secondary phloem and secondary xylem are derived from the

(A) root apical meristem
(B) shoot apical meristem
(C) vascular cambium
(D) cork cambium
(E) cotyledons

51. Protheria are egg-laying animals that are classed as mammals because they possess

(A) hair over parts of the body
(B) a hypothalamus mechanism that regulates their body temperature
(C) a four-chambered heart
(D) a single camera-type eye
(E) a vertebral column

52. A successful blood transfusion requires that the recipient have

(A) type O blood with an Rh factor that is positive
(B) different antibodies and antigens than those of the donor
(C) antibodies against the red blood cells of the donor
(D) a red blood cell count that is identical to that of the donor's blood
(E) the same antigens and antibodies as those in the donor's blood

53. The stage of mitosis during which the chromosomes coil up, the nuclear membrane disappears, and the centrioles migrate to the poles is

(A) prophase
(B) metaphase
(C) anaphase
(D) telophase
(E) interphase

54. All tracheophytes possess

(A) gametes that exhibit vascular tissues
(B) reproductive structures called flowers
(C) a dominant gametophyte generation
(D) a dominant sporophyte generation
(E) separate sexes in the sporophyte generations

55. In the life cycle of slime molds, free-living unicellular individuals come together to form a "slug" that is capable of movement. This life cycle change is useful in the study of

(A) tissue formation
(B) mitosis
(C) meiosis
(D) alternation of generations
(E) fertilization

56. The significance of mitosis is

(A) diploid cells produce haploid cells for sexual reproduction

(B) it serves as a mechanism for the production of gametes
(C) it provides a time for synapsis and crossing over
(D) diploid cells produce diploid cells that are copies of the parent cells
(E) haploid cells combine to form diploid cells

57. Variable-shaped, thick-walled, dead cells that strengthen xylem are
(A) sclerid cells
(B) vessel cells
(C) parenchyma cells
(D) mesophyll cells
(E) none of these

58. Auxin (IAA) is a plant hormone that can be characterized as a/an
(A) stimulator of cell elongation in stems
(B) inhibitor of cell elongation in the cambium
(C) both (A and B)
(D) neither (A nor B)
(E) stimulator of cellulose production

59. The fluid–mosaic model proposes that the membrane of a cell is a structure with
(A) two layers of protein with a layer of lipid on either side
(B) a double layer of lipids that possess globular proteins of various sizes
(C) a double layer of lipids with a layer of protein on either side
(D) two lipid layers sandwiched between two carbohydrate layers
(E) none of these

Questions 60 and 61 refer to the graph below.

60. The conclusion that can be drawn from the graph is
(A) there is an inverse proportionality between substrate availability and product concentration
(B) as substrate is changed into product, substrate availability decreases
(C) there is no relationship between substrate and product
(D) substrate reaches maximum availability after 25 minutes
(E) all enzymes are product dependent

61. The time when substrate and product availability are about equal is after
(A) 0 minutes
(B) 12 minutes
(C) 25 minutes
(D) 41 minutes
(E) 50 minutes

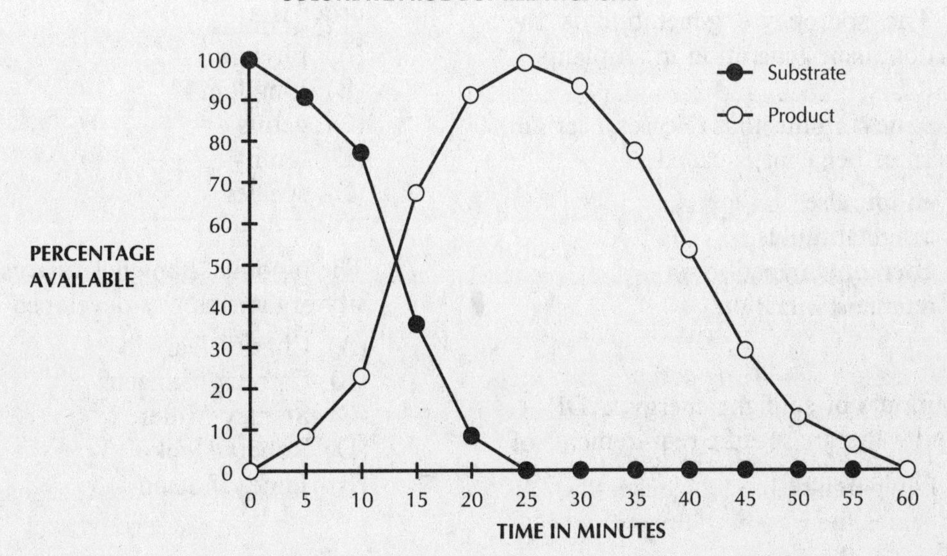

62. A substance that has a pH of 6.3 would be expected to be a
    (A) strong acid with a sour taste
    (B) weak acid with a sour taste
    (C) weak acid with a sweet taste
    (D) weak base with a sour taste
    (E) strong acid with a sweet taste

63. In some flowers, the fruit develops from the
    (A) petiole
    (B) recepticle
    (C) ovule
    (D) endosperm
    (E) plumule

64. Brown algae belong to the phylum
    (A) Chordata
    (B) Eumycophyta
    (C) Sarcodina
    (D) Phaeophyta
    (E) Ciliophora

65. Select the statement that is true for sexually reproducing plants.
    (A) Gametophytes produce gametes by means of meiosis.
    (B) Gametophytes produce gametes by means of mitosis.
    (C) Sporophytes and gametophytes both contain diploid cells.
    (D) Sporophytes and gametophytes both contain haploid cells.
    (E) The sporophyte generation is the dominant generation in all plants.

66. A specific stimulus that releases a certain pattern of behavior is called
    (A) an impulse
    (B) a sign stimulus
    (C) afferent summation
    (D) tetanic summation
    (E) a reflex

67. The inputs of sunlight energy, ADP + P, and water are specific requirements of
    (A) photosystem I
    (B) photosystem II
    (C) glycolysis
    (D) electron transport system
    (E) Benson–Calvin Cycle

68. The simplest type of learning is
    (A) habituation
    (B) conditioned reflex
    (C) trial and error
    (D) latent learning
    (E) insight learning

69. Substances secreted by one individual that stimulate a behavior response in another individual of the same species are known as
    (A) hormones
    (B) pheromones
    (C) catacholeamines
    (D) neurotransmitters
    (E) inducers

70. A fruit is defined as that part of the plant that
    (A) is the only part eaten
    (B) is rich and lush all of the time
    (C) contains the seeds of the plant
    (D) can be peeled and stored for extended periods of time
    (E) grows in response to sunlight

71. The most immediate subdivision of an order is a
    (A) phylum
    (B) kingdom
    (C) genus
    (D) family
    (E) species

72. The binomial nomenclature used to classify organisms was developed by
    (A) Charles Darwin
    (B) Carolus Linnaeus
    (C) Stanley Miller
    (D) Robert Hooke
    (E) James Watson

73. The behavioral state in which homeothermic animals seek to regulate body temperature is
(A) sleep
(B) curiosity
(C) thirst
(D) mating
(E) sweating

74. The plant kingdom includes members of the phyla
(A) Coniferophyta, Chlorophyta, and Pyrrophyta
(B) Lycopodophyta, Bryophyta, and Anthophyta
(C) Mollusca, Arthropoda, and Annelida
(D) Chrosophyta, Sporozoa, and Chordata
(E) Myxomycophyta, Eumycophyta, and Rhodophyta

75. The biome which is characterized by conifer forests, long, severe winters, and a permanent cover of snow is the
(A) chaparral
(B) temperate grasslands
(C) temperate deciduous forest
(D) tundra
(E) taiga

76. Which of the following changes does *not* affect the rate of enzyme reaction?
(A) temperature
(B) pH
(C) substrate concentration
(D) enzyme concentration
(E) none of these

77. Fatty acids and glycerol are to simple lipids as
(A) monosaccharides are to polypeptides
(B) amino acids are to proteins
(C) glucose is to glycerol
(D) adenine is to thymine
(E) nucleotides are to amino acids

78. Nephridia are
(A) excretory organs characteristic of the phylum Annelida
(B) excretory organs characteristic of the class Chondrichthyes
(C) the fundamental units of the central nervous system
(D) the specialized respiratory structures found in insects
(E) cell organelles which contain powerful proteolytic enzymes

79. Bundles of column-shaped cells that help in the support of soft stems, xylem, and phloem are made up of
(A) collenchyma cells
(B) parenchyma cells
(C) suberin
(D) companion cells
(E) lenticels

80. After the formation of the zygote in angiosperm reproduction, many mitoses lead to the development of a/an
(A) series of flowers without sepals
(B) endosperm nucleus
(C) ovule
(D) embryo with stored food and a seed coat
(E) anther with numerous pollen grains

81. The modification of the cell that provides single-celled organisms like paramecium with the ability to move freely is the
(A) cilium
(B) endocytic vescicles
(C) microvillus
(D) nucleus
(E) nucleolus

82. The diploid cell resulting from the fusion of two haploid cells during the process of fertilization is the
(A) coenocyte
(B) syncytium
(C) daughter cells
(D) zygote
(E) morula

83. Mitochondrion is to energy production as

(A) mitochondrion is to cristae
(B) lysosome is to intracellular digestion
(C) mitochondrion is to nuclear membrane
(D) lysosome is to extracellular digestion
(E) cell is to tissue

Questions 84 and 85 refer to the line graph below that depicts the results of blood estrogen measurements in women living under different conditions of crowding.

84. According to the graph, moderately crowded females show the highest level of blood estrogen on day

(A) 28
(B) 10
(C) 14
(D) 24
(E) 21

85. The most reasonable conclusion that can be drawn from the data presented in the graph is that

(A) moderate crowding results in an elevated blood estrogen level without disturbing the fluctuations over time seen in uncrowded women
(B) severe crowding results in an elevated blood estrogen level without disturbing the fluctuations over time seen in uncrowded women
(C) both moderate and severe crowding reduce blood estrogen levels
(D) the estrogen level on day 10 under severe crowding conditions is higher than under moderate crowding conditions
(E) levels of estrogen are independent of crowding conditions

86. The anterior modification of the sclera in the human eye is the

(A) retina
(B) pupil
(C) iris
(D) cornea
(E) vitreous humor

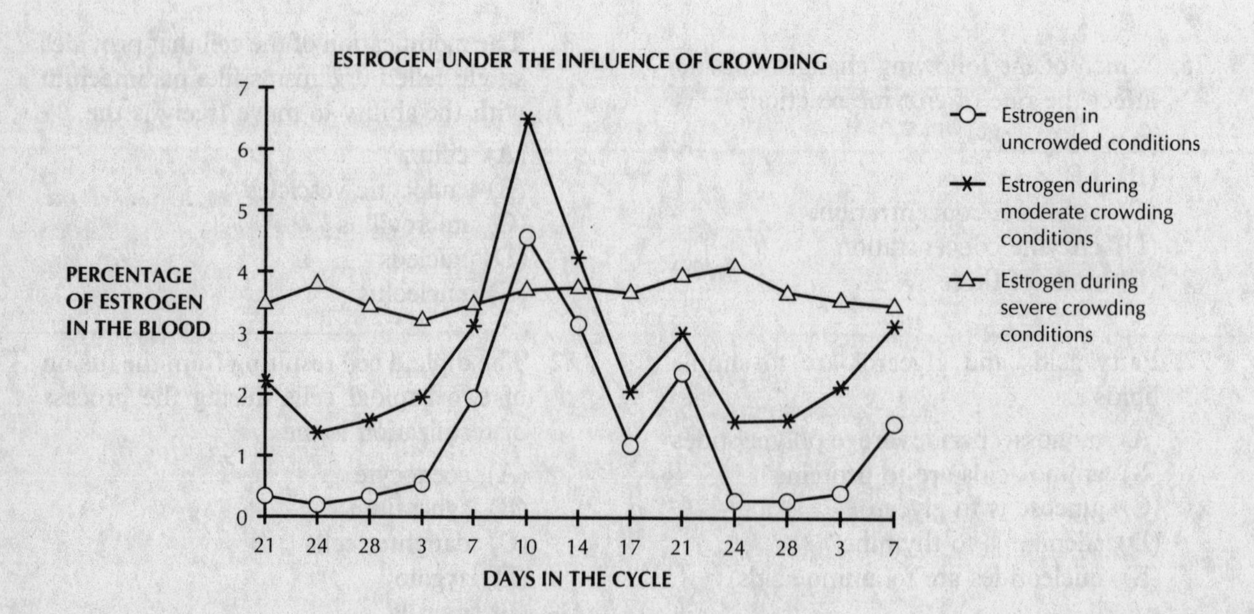

NUTRITIONAL DATA SUPPLIED ON THE SIDE OF A POPULAR CEREAL BOX

	1 Cup Cereal	1 Cup Cereal + 8-oz Skimmed Milk	1 Cup Cereal + 8-oz Whole Milk
Calories	110	150	180
Protein (grams)	3	7	7
Carbohydrate (grams)	31	37	37
Fats (grams)	1	1	5
		Cereal	Cereal + Milk
Starch and Relative Carbohydrates (grams)		15	15
Sucrose and Other Sugars (grams)		12	18
Dietary Fibers (grams)		4	4
Total Carbohydrates (grams)		31	37

Questions 87–90 are based on the table above.

87. From the available data one can conclude that
    (A) a cup of whole milk has 70 more calories than a cup of skimmed milk
    (B) two cups of skimmed milk have more calories than one cup of whole milk
    (C) one cup of skimmed milk has more calories than one cup of whole milk
    (D) one cup of whole milk has 15 grams of starch and relative carbohydrate
    (E) one cup of whole milk has 4 grams of dietary fiber

88. Another conclusion to be drawn from the available data is
    (A) whole milk has more protein than skimmed milk
    (B) one cup of skimmed milk has 1 gram of fat
    (C) one cup of whole milk has 6 grams of sucrose and other sugars
    (D) one cup of whole milk has 15 grams of starch and relative carbohydrate
    (E) skimmed milk has more fat than the cereal

89. The difference in total carbohydrate between cereal and cereal + milk comes from
    (A) starch and relative carbohydrate
    (B) sucrose and other sugars
    (C) the fat in the whole milk
    (D) the proteins in the whole milk
    (E) dietary fibers in the milk

90. The available data leads to the conclusion that
    (A) cereal + whole milk has more protein than cereal + skimmed milk
    (B) cereal + whole milk is less fattening than cereal + skimmed milk
    (C) cereal alone is less fattening than cereal + milk
    (D) cereal + whole milk provides less energy than cereal + skimmed milk
    (E) cereal + whole milk provides more vitamins than cereal + skimmed milk

***Directions:* Each group of questions below consists of five lettered headings followed by a list of numbered phrases or sentences. For each numbered phrase or sentence, select the one heading which is most closely related to it. One heading may be used once, more than once, or not at all in each group.**

Questions 91–96.

(A) Morlula
(B) Gastrulation
(C) Zygote
(D) Seminal fluid
(E) Clitoris

91. The female analog of the penis.

92. Produced by both the prostate gland and the Cowper's gland.

93. An early stage of development in humans and other mammals.

94. Produced by the male reproductive tract.

95. The fluid component of an ejaculate.

96. The diploid cell resulting from the fusion of an egg cell and sperm cell.

Questions 97–99.

(A) Divergent evolution
(B) Environmental adaptation
(C) Convergent evolution
(D) Physiological adaptation
(E) Parallel evolution

97. The evolution of marsupials in Australia and the evolution of placental mammals in other parts of the world.

98. A state of adjustment to the environment.

99. The evolution of brown bears and polar bears.

Questions 100–103.

A pure breed of mice are all derived from a single male and female. 53 males and 59 females were found to have blood type N while a single male exhibited a new blood type designated type R. A set of test matings involving this male resulted in the following data:

P_1 R male × N female

F_1

19 R males 16 N males 31 RN females

100. The P_1 male does not show blood type N because
(A) there is a blocking gene on the y-chromosome
(B) the male lacks DNA on the x-chromosome
(C) the gene penetrance for R is variable
(D) the R gene is sex-linked
(E) the R gene is a recessive to the N gene

101. It is clear from the data that the
(A) y-chromosome does not have a locus for blood type in mice
(B) x-chromosome does not have a locus for blood type in mice
(C) x-chromosome has two separate loci for N and R blood type genes
(D) somatic chromosomes play an important role in mouse blood type
(E) appearance of blood type R was the result of mutagens in the mouse food

102. One possible conclusion that can be drawn from the available data is that the
(A) N gene is dominant over the R gene
(B) N and R genes are codominant
(C) N gene is recessive to the R gene
(D) N and R genes are part of a multiple allelic series
(E) genes for N and R are linked

103. The data strongly supports the idea that
   (A) the gene for R resulted from a spontaneous mutation of the N gene
   (B) males are more likely to express the R gene than are females
   (C) the gene for R is on a somatic chromosome
   (D) mutations are usually to a recessive condition
   (E) acquired characteristics are passed on to the next generation

Questions 104–108.

(A) Bile pigment
(B) Trypsin
(C) Pepsin
(D) Chyme
(E) Bile salts

104. The protein digesting substance that is a component of gastric juice.

105. The mixed substance that is periodically moved through the pyloric sphincter into the duodenum.

106. The substance that hydrolyses proteins in an acidic environment.

107. The substance that confers a brown color to the feces.

108. The waste substance produced by the liver during the breakdown of worn out erythrocytes.

Questions 109–112.

(A) Biotic Community
(B) Ecosystem
(C) Trophic level
(D) Climax Community
(E) Tundra

109. The relatively stable body in an ecological succession.

110. A form of grassland that is characterized by a permafrost, a layer of permanently frozen subsoil.

111. Secondary consumers.

112. All interacting populations inhabiting a common environment.

GO ON TO THE NEXT PAGE.

Questions 113–116. Refer to Figures 2-1 and 2-2.

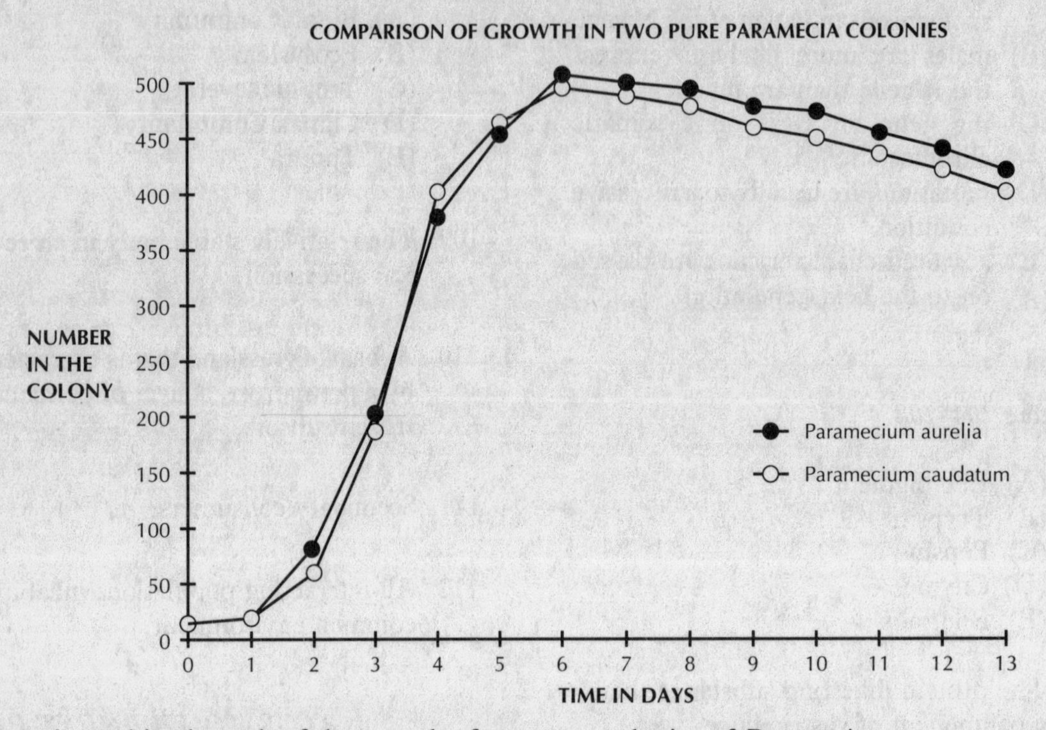

Fig. 2-1. A combined graph of the growth of two pure colonies of *Paramecia*.

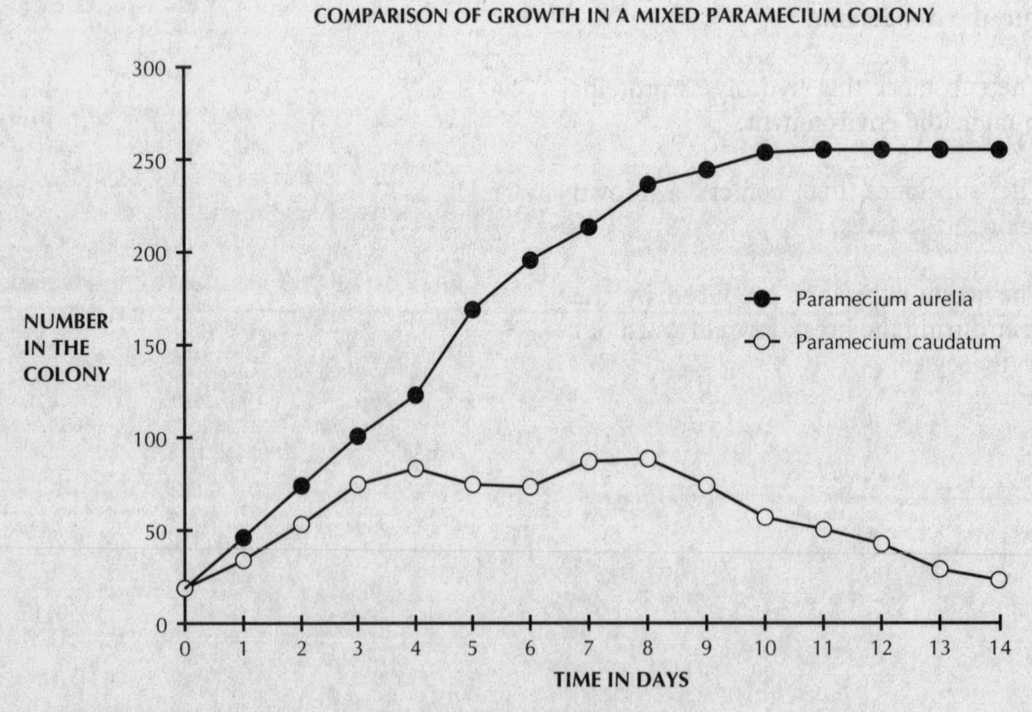

Fig. 2-2. A graph of the growth of a mixed colony of *Paramecia*.

113. In pure culture, *Paramecium caudatum* attains 50% maximum population size at:
(A) 1–2 days
(B) 3–4 days
(C) 6–7 days
(D) 10–11 days
(E) 13–14 days

114. From these graphs one can see that
(A) *Paramecium caudatum* dies out in the presence of *Paramecium aurelia*
(B) *Paramecium aurelia* feeds on *Paramecium caudatum*
(C) *Paramecium aurelia* reach sexual maturity much earlier than *Paramecium caudatum*
(D) *Paramecium caudatum* in pure culture will die out after 14 days
(E) *Paramecium caudatum* is better adapted to interspecific competition than *Paramecium aurelia*

115. In mixed populations
(A) *Paramecium caudatum's* growth is unaffected while *Paramecium aurelia's* is diminished
(B) *Paramecium aurelia's* growth is unaffected while *Paramecium caudatum's* is diminished
(C) both *Paramecium aurelia* and *Paramecium caudatum* are unaffected
(D) both *Paramecium aurelia* and *Paramecium caudatum* are affected
(E) *Paramecium aurelia* initially grows faster than *Paramecium caudatum*

116. From these two graphs one can conclude that
(A) in pure culture, *Paramecium caudatum* grows better than *Paramecium aurelia*
(B) in pure culture, *Paramecium aurelia* grows better than *Paramecium caudatum*
(C) in mixed culture, *Paramecium caudatum* competes more effectively than *Paramecium aurelia*
(D) in mixed culture, *Paramecium aurelia* competes more effectively than *Paramecium caudatum*
(E) mixed populations survive longer than pure populations

Questions 117–120.

(A) Induction
(B) Differentiation
(C) Archenteron
(D) Cleavage
(E) Neurulation

117. The primitive gut cavity produced during gastrulation.

118. Cell division without increase in mass.

119. The process that helps explain the orderly development of the embryo.

120. The process leading to the formation of the morula.

SECTION 2

Time—90 minutes

Percent of Total Grade—40

The questions in this section are mandatory. Answer *ALL* four questions.

1. Name and describe the major reactions and sites of reactions of photosynthesis. State the relationships of these reactions. Include a definition and brief discussion of C_3 and C_4 plants.
2. The human alimentary tract consists of a number of specialized structures and modifications.
   a. In sequence, describe these structures and their modifications.
   b. In sequence, discuss enzymatic digestion of proteins in the human alimentary tract.
3. The process of evolution is a concept that explains the history and diversity of life on earth.
   a. Define evolution in modern terms.
   b. List and briefly discuss five different lines of evidence that support evolution.
4. Nutrients cycle through ecosystems.
   a. Describe the basic components of an ecosystem.
   b. Using the complete list below, construct a food web. Indicate the trophic level for each.

BRUSH VEGETATION	ROBIN
CARIBOU	SOIL BACTERIA AND FUNGUS
EAGLE	SPIDER
HERBIVOROUS INSECT	WOLF

ANSWER KEY

SECTION I

1. **E**
2. **B**
3. **B**
4. **A**
5. **D**
6. **A**
7. **D**
8. **B**
9. **C**
10. **E**
11. **B**
12. **E**
13. **B**
14. **B**
15. **D**
16. **D**
17. **C**
18. **C**
19. **B**
20. **B**
21. **A**
22. **B**
23. **B**
24. **C**
25. **C**
26. **A**
27. **B**
28. **A**
29. **C**
30. **A**
31. **B**
32. **A**
33. **D**
34. **E**
35. **B**
36. **D**
37. **A**
38. **C**
39. **C**
40. **B**
41. **B**
42. **A**
43. **C**
44. **D**
45. **C**
46. **B**
47. **B**
48. **D**
49. **C**
50. **C**
51. **A**
52. **E**
53. **A**
54. **D**
55. **A**
56. **D**
57. **A**
58. **C**
59. **B**
60. **B**
61. **B**
62. **B**
63. **B**
64. **D**
65. **B**
66. **B**
67. **B**
68. **A**
69. **B**
70. **C**
71. **D**
72. **B**
73. **E**
74. **B**
75. **E**
76. **E**
77. **B**
78. **A**
79. **A**
80. **D**
81. **A**
82. **D**
83. **B**
84. **B**
85. **A**
86. **D**
87. **B**
88. **C**
89. **B**
90. **C**
91. **E**
92. **D**
93. **A**
94. **D**
95. **D**
96. **C**
97. **E**
98. **B**
99. **A**
100. **D**
101. **A**
102. **B**
103. **A**
104. **C**
105. **D**
106. **C**
107. **A**
108. **A**
109. **D**
110. **E**
111. **C**
112. **A**
113. **B**
114. **A**
115. **D**
116. **D**
117. **C**
118. **D**
119. **A**
120. **B**

DIAGNOSTIC CHART

SECTION 1

This chart will provide you with the opportunity to identify those areas of biology in which you have done well and those areas in biology in which you need to improve your knowledge and understanding. For each correct answer, place an X in the space provided. You are doing well if you have 80–90% of the answers correct in each topic area. You are also doing well if your *total* score is 85% or better.

TOPIC	QUESTIONS ON THE EXAMINATION	NUMBER CORRECT
Basic Chemistry	16 17 28 43 62 77	______
Cell Structure and Function	25 59 81 83	______
Enzyme Structure and Function	18 31 32 35 76	______
Cell Division	27 45 53 56	______
Energy Transforms in Living Systems	1 14 15 26 33 67	______
Origin of Life	6 21 23 37	______
Taxonomy	51 55 64 71 72 74	______
Higher Plants: Structure and Function	24 34 44 50 57 58 63 79	______
Higher Plants: Reproduction and Development	2 54 65 70 80	______
Higher Animals: Structure and Function	38 47 52 78 86 104 105 106 107 108	______
Higher Animals: Reproduction and Development	82 91 92 93 94 95 96 117 118 119 120	______
Heredity	3 19 46 100 101 102 103	______

Category	Questions	Score
Evolution	7 12 20 22 36 40 48 97 98 99	____
Ecology	5 8 11 49 75 109 110 111 112	____
Behavior	13 39 66 68 69 73	____
Graph, Chart or Table Interpretation	4 5 9 10 29 30 41 42 60 61 84	
	85 87 88 89 90 113 114 115 116	____
Total Correct		____

EXPLANATORY ANSWERS

SECTION 1

1. **(E)** is correct. (A and D) are correct if the statements are reversed. (B) is correct if molecular oxygen is replaced by $NADPH_2$. (C) is always an anaerobic process. If there is a lack of molecular oxygen, the pyruvic acid will, for a time, move into the fermentation pathway rather than the Krebs Cycle pathway.

2. **(B)** is correct. (A) forms the ring of modified flower leaves within the sepals. These leaves often bear color and patterns that attract animal pollinators. (C) are the modified flower leaf structures that bear the male reproductive organs. (D) are the modified flower leaf structures that bear the female reproductive organs. (E) are the pollen receptor segments of the carpels.

3. **(B)** is the best of current hypotheses. (A) is nonsense. (C) is unrelated to the question. (D and E) are not supported by the current research.

4. **(A)** is correct. (B) is nonsense. (C) has nothing to do with the available information. (D and E) are opposite to the data presented.

5. **(D)** is correct. Initiation and continuation of legalized hunting would have the initial effect of somewhat reducing the population and would contribute to the maintenance of a reduced population.

6. **(A)** is correct. (B) are complex carbohydrates composed of long chains of glucose. (C) Only an entire cell can accomplish this. (D) are nucleic acids.

7. **(D)** is correct. (A) are noted for their developing a model for DNA structure. (B) is noted for his theory of organic evolution through natural selection and survival of the fittest. (C) is known as the father of genetics. (E) is the first one who coined the term "cell."

8. **(B)** is correct. (A, C, D, and E) are secondary, tertiary, or quaternary consumers.

9. **(C)** is correct. It is the only curve that has a 70% saturation of hemoglobin at a partial pressure of 40 mmHg.

10. **(E)** is correct. As stated below the graph, an increasing acidity leads to a reduced percentage of oxygen saturation.

11. **(B)** are cycles that have a 24-hour period that affect many physiological processes.

12. **(E)** is correct. (A and B) both speculate about the inorganic origins of life. They never conducted any related experiments. (C) speculated about the organic evolution of life forms. (D) demonstrated that dry heating of amino acids could yield long chain proteinoids.

13. **(B)** is correct. Experiments with homing pigeons have demonstrated that under conditions of cloud cover, small battery-powered uniform magnetic fields distorted the pigeons' ability to find their way home. On clear days, the magnetic fields did not appear to have a significant influence on their homing abilities. It appears that navigating by means of magnetic lines of force serves as a backup to solar navigation in pigeons.

14. **(B)** is correct. (A and C) are both predicted by the first law of thermodynamics. (D) It is believed that black holes capture all energy that comes near. (E) would be in conflict with the second law of thermodynamics.

15. **(D)** is the dark reaction, a light-independent chemistry that produces glucose from carbon dioxide, hydrogen from water, and ATP energy derived from sunlight. (A) is the cellular respiration process that degrades pyruvic acid to carbon dioxide while releasing the energy necessary to produce ATP in the electron transport system. (B) is the cellular respiration process that degrades glucose to pyruvic acid with some production of ATP. (C) is the

photosynthetic process which fixes energy-poor carbon dioxide into energy-rich glucose. (E) is the cellular respiration process that produces ATP from energy that is derived from glycolysis and the Krebs Cycle.

16. **(D)** is commonly called "animal starch." (A, B, and C) are all monosaccharides. (E) is a plant polysaccharide.

17. **(C)** is capable of forming straight chains, branched chains, and/or rings. (A, D and E) are important elements but are not capable of serving as the structural bases for organic compounds. (B) is a compound and not an element.

18. **(C)** are considered coenzymes and are necessary for the function of enzymes that lack prosthetic groups. (A) would be true of enzymes with prosthetic groups. (B) is nonsense as different enzymes work at different pH values. (D) enzymes are usually quite specific.

19. **(B)** is correct. (A and D) are diseases of the nervous system. (C) is a disease of the immune system. (E) is a genetic deficiency disease.

20. **(B)** Bernard Kettleworth in England noticed that collections of peppered-moths caught between the late 1800s to the early 1900s shifted from being composed primarily of light-colored moths to primarily dark-colored moths. The Industrial Revolution caused darkening of tree trunks making the light-colored moths more subject to predation than the dark-colored moths.

21. **(A)** is correct. (B and C) are nonsense. (D) Heterotrophs are believed to have evolved before autotrophs.

22. **(B)** is correct. (A) describes the situation where two ancestral species resemble one another's adaptation and the descendents of each species evolve along similar lines. (C) describes the situation where one population of a species becomes isolated from other populations of the same species. Then a different course of evolution results from selective pressures that are different from those affecting other populations of that species. (D) is the result of populations of different species being exposed to similar selective pressures and exhibiting similar adaptations. (E) do not apply to this situation.

23. **(B)** is correct. The primitive atmosphere of the earth is thought to have been composed primarily of ammonia gas, methane gas, hydrogen gas, and water vapor. This reducing atmosphere was changed to an oxidizing atmosphere with the development of photosynthesis.

24. **(C)** is correct. Dicot leaves have both a spongy and palisade mesophyll. (A) is characterized by scattered vascular bundles and lacks a pith. (B) is characterized by annual rings of xylem. (D) is characterized by a complete spongy mesophyll. (E) is characterized by a broken endodermis and lacks a pith. The center is filled with xylem.

25. **(C)** is correct as lysosomes are the only known cell organelles to contain powerful proteolytic enzymes but no DNA.

26. **(A)** are chlorophyll-containing cell organelles in which photosynthesis occurs. (B) are coiled strands of DNA and their associated proteins. They are only observed during mitosis or meiosis. (C) are organelles in which oxidative phosphorylation occurs. (D) are organelles that are sites of protein synthesis. (E) are organelles that are not associated with photosynthesis.

27. **(B)** is correct. (A) occurs during interphase and prophase, respectively. (C) occurs during synapsis in metaphase. (D) is characteristic of telophase in animal cells. (E) occurs during the synthesis phase of the interphase.

28. **(A)** is correct. (B) is the secondary structure of protein. (C) is the tertiary structure of protein. (D) is the quaternary structure of protein. (E) is not related to proteins.

29. **(C)** is correct.

30. **(A)** is correct. (B) There is no information on the degree or rate of development. (C) is the reverse of (A). (D and E) are not reflected in the chart.

31. **(B)** is correct. All enzymes are tertiary proteins. (A) are the building blocks of nucleic acids. (C) are the building blocks of simple lipids. (D) are single and double sugars. (E) relate the citric acid cycle and the nervous system, respectively.

32. **(A)** is correct. (B, C, and D) do not make any sense.

33. **(D)** is correct. Noncyclic photophosphorylation splits water and captures sunlight energy to make hydrogen and energy available for the Calvin Cycle. (A) is aerobic and converts glucose to pyruvic acid with some production of ATP. (B) requires molecular oxygen to pick up hydrogen ions coming off the electron transport system. (C) captures sunlight energy and stores it in the high-energy bonds of ATP. (E) is a catabolic series of events that require the presence of molecular oxygen.

34. **(E)** is the generalization while the other answers are specific. (A) is the function of auxins and gibberellins. (B) is the function of auxins and cytokinins. (C) is the function of phytochrome and abscisic acid. (D) is the function of abscisic acid.

35. **(B)** is correct. (A) would prevent the substrate complex bond from breaking. (C and D) are both nonsense.

36. **(D)** This is a concept that is contrary to the one proposed by Darwin. This concept was originally proposed by Lamarck.

37. **(A)** is correct. (B, C, D, and E) were thought to be evolutionary advancements that occurred after the development of the first cells.

38. **(C)** is correct. (A) is a protein that is associated with DNA. (B) is a multinucleate plant cell. (D) is the structure that holds sister chromatids together during cell division. (E) is the building block of nucleic acids.

39. **(C)** is correct. (A) is a response to a normally neutral stimulus that has become associated with some type of expected behavior. (B) is a random response with either a reward or penalty that is associated with the behavior. (D) is learning that occurs without any obvious behavior at the time that the learning occurs. (E) is the simplest form of learning in which the magnitude of the behavior response gradually lowers with acclimation to a particular stimulus.

40. **(B)** is the correct answer. The twelve species of finches that Darwin had observed had adaptations of the beak that reflected their adaptations to a wide variety of diets.

41. **(B)** is correct. After four days of exposure to 7 ppm nitrogen dioxide there are no additional deaths, while there continue to be deaths in the 35 ppm and 50 ppm nitrogen dioxide exposed groups.

42. **(A)** is correct. As the concentration of nitrogen dioxide increases above 7 ppm there is an increased mortality.

43. **(C)** is correct. The atomic number 14 indicates that there are 14 protons in the atom. The atomic weight indicates that the sum of protons and neutrons is 29. Therefore, there are 15 neutrons in the atom. As atoms are electrically stable, there are then 14 electrons in the atom. (A and B) are both reversed. (D) is incorrect as Ne is inert. (E); see (C).

44. **(D)** Auxin causes cell elongation in stems. The cells of the side with the agar block will elongate faster than the cells on the opposite side. The tip therefore bends away from the side of the greatest cell elongation.

45. **(C)** is correct.

46. **(B)** is correct. Metabolic diseases are usually the result of an enzyme deficiency. The lack of an enzyme in some critical pathway prevents the synthesis of some vital product. If metabolic diseases were the result of dominant or codominant genes rather than recessive genes, then a large number of individuals would exhibit metabolic disorders.

47. **(B)** is correct. These portions of the nephron are in the renal cortex while the limbs, the

loop of Henle, and collecting ducts are generally found in the renal medulla. (C) characterizes the liver. (D) characterizes the gastric submucosa. (E) characterizes the pancreas.

48. **(D)** is correct. (A) first expressed the concept of biogenesis, life comes from life. (B) presented convincing evidence that blood circulates around the human body. (C) along with Crick described the structure of DNA. (E) proposed the theory of evolution through natural selection.

49. **(C)** is correct. Ammonification results from bacterial and fungal decomposition of organic materials. This decomposition results in the production of ammonia. Nitrification results from bacteria that convert ammonia into nitrites and others that convert nitrites into nitrates. Assimilation results in the utilization of nitrates for the synthesis of amino acids and nucleosides.

50. **(C)** is correct. (A) gives rise to primary growth in the root. (B) gives rise to primary growth in shoots. (D) gives rise to the growth of new cork in the older dicot stem. (E) is the seed-leaf found in flowering plants.

51. **(A)** plus mammary glands characterize the mammals. (B) Both birds and mammals and even a few fish can regulate body temperature. (C) is typical of both birds and mammals. (D) is found among some of the invertebrates. (E) is common to fishes, amphibians, reptiles, birds, and mammals.

52. **(E)** is correct. (A and D) are both nonsense. (B and C) will cause erytyhrocytes to agglutinate.

53. **(A)** is correct. (B) is a time of spindle formation and chromosome lining up along the equatorial plane. (C) is a time of chromatid pair separation. (D) is a time of cytoplasmic division and the establishment of new nuclei in each daughter cell. (E) is a time of cell growth and DNA replication.

54. **(D)** Tracheophytes (ferns, gymnosperm, and angiosperms) all have a dominant sporophyte generation. (A) None of the gametophytes have vascularized. (B) angiosperms produce flowers for reproduction while gymnosperms produce cones for reproduction.

55. **(A)** is correct. Tissues are aggregates of similar cells with similar function. (B and C) No form of cell division is involved in the life cycle described. (D) There is no change in ploidy in the part of the life cycle described. (E) There are no gametes present.

56. **(D)** is correct. (A, B, and C) are true of meiosis. (E) is the process of fertilization.

57. **(A)** is correct. (B) are short, wide, thick-walled, dead cells that conduct water and minerals. (C) are relatively large, thin-walled, relatively unspecialized living cells that are capable of starch storage and the lateral conduction of water. (D) are relatively thin-walled photosynthetic cells found between the upper and lower epidermis of leaves.

58. **(C)** is correct. (E) is a function of ethylene.

59. **(B)** is correct. The model proposed is a "sea of lipid" in which tertiary protein molecules float like "icebergs." (A and D) are fancies of imagination. (C) is an old model.

60. **(B)** is correct. During the first 25 days of the experiment, the substrate availability goes from 100% to 0% while the product availability goes from 0% to 100%.

61. **(B)** is correct. Substrate and Product are equal when each has a 50% availability.

62. **(B)** is correct. Acids have a sour taste. Anything that has a pH of less than 7.0 is considered to be an acid. Weak acids have a pH close to 7.0 while strong acids have a pH close to 1.0.

63. **(B)** is correct. (A) is the narrow part of the leaf that attaches the blade to the stem. (C) is the flower structure in which the diploid megaspore mother cell develops into four haploid megaspores (eggs). (D) is the triploid nutrient tissue used by the embryo while it grows in the seed.

64. **(D)** is correct. (A) includes all higher animals with some form of internal support. (B) includes molds, yeasts, mushrooms, and ringwork fungus. (C) includes the amoebas. (E) includes the ciliates and suctorians.

65. **(B)** is correct.

66. **(B)** is correct. (A) is a wave of membrane depolarization in the neuron. (C and D) are nonsense. (E) is an inborn autonomic response to a stimulus. It is dependent on the existence of a fixed neural pathway.

67. **(B)** is correct. (A) requires sunlight energy and ADP + P. (C) requires glucose and ADP + P. (D) requires $NADPH_2$, ADP + P, and molecular oxygen. (E) requires ADP, $NADPH_2$, and CO_2.

68. **(A)** is correct. (B) is a response to a normally neutral stimulus. The stimulus has become associated with some expected behavior. (C) is a random response with either a reward or penalty associated with the behavior. (D) is a response without any obvious behavior. (E) requires the integration of past experiences in arriving at a novel behavior in the solution of a novel problem.

69. **(B)** is correct. (A) are chemicals produced by one group of cells, a tissue, or an organ that have a regulatory function on another group of cells, a tissue, or an organ in another part of the body. (C and D) are chemicals that are found in the nervous system. (E) are substances that play important roles in embryonic development.

70. **(C)** is correct. (A) many parts of plants can be eaten (roots, stems, and leaves). (B, D, and E) are nonsense.

71. **(D)** is correct. (A) is the usual primary subdivision of a kingdom. (B) is any of five major categories into which organisms can be classified. (C) is the usual primary subdivision of a family. (E) is the usual primary subdivision of a genus.

72. **(B)** is correct. (A) is noted for his theory of organic evolution by means of natural selection with the survival of the fittest. (C) experimentally demonstrated that organic molecules could be spontaneously generated from simple inorganic compounds in the presence of heat and electrical discharge. (D) is noted for having coined the term "cell." (E) along with Frances Crick determined the structure of DNA.

73. **(E)** is correct. (A) is a higher animal condition in which the organism is active and responsive to stimuli. (B) is a higher animal condition that may well be a complex of responsive behaviors under a single name. (C) is a land animal condition in which the animal seeks water. (D) is a condition that appears to be determined by innate factors rigidly controlled by the nervous system in lower animals and higher centers of the central nervous system in higher animals.

74. **(B)** is correct. (A) Pyrrophyta belong to the Kingdom Protista. (C) All belong to the Kingdom Animalia. (D) Chrysophyta and Sporozoa are protists. Chordates are higher animals. (E) Myxomycophyta and Eumycophyta belong to the Kingdom Fungi.

75. **(E)** is correct. (A) is a region dominated by small trees and/or spiny shrubs, and is subject to mild, rainy winters alternating with dry, hot summers. (B) is a region of rolling flat terrain with periodic draughts and hot and cole seasons. (C) is a region with a warm growing season and moderate precipitation alternating with cold periods poorly suited for growth. (D) is a region of grassland characterized by a permanently frozen subsoil.

76. **(E)** is correct. (A, B, C, and D) are all changes that influence the rate of enzyme reaction.

77. **(B)** is correct. The relationship is "the building block to the complex molecule." (A) Monosaccharides are building blocks of polysaccharides. Amino acids are building blocks of polypeptides. (C) Glucose is a carbohydrate while glycerol is part of a lipid. (D) Adenine and thymine are organic bases in nucleic acids. (E) Nucleotides are the building blocks of nucleic acids. Amino acids are the building blocks of proteins.

78. **(A)** is correct. (B) is the primitive kidney. (C) is the neuron. (D) are the tracheal tubes. (E) are the lysosomes.

79. **(A)** is correct. (B) is composed of polyhedral cells used for food storage. (C) is the waxy waterproofing material found on the outside of the bark. (D) are the small nucleated cells that nurture the anucleate sieve tubes in the phloem. (E) are the longitudinal cracks in the bark of the stem. They allow for gas exchange in the stem tissues.

80. **(D)** is correct. (A, B, C, and E) are nonsense.

81. **(A)** is the cellular organelle which contains an arrangement of microtubules that allow it to provide a whipping action. (B) is a cell inclusion, not a cell organelle. (C) is a folding of the cell membrane. It allows the membrane to have an increased surface area. (D) is not a cellular modification for locomotion. (E) is a diffuse structure in the nucleus.

82. **(D)** is correct. (A) is a multinucleate plant cell. (B) is a multinucleate animal cell. (C) is one of the two diploid cells produced during mitosis. (E) is the ball of cells formed by numerous mitoses after fertilization.

83. **(B)** is correct. The relationship is cell organelle to organelle function. (A) is structure to structure. (C) is structure to different structure. (D) is structure to different function. (E) is structure to level of structure.

84. **(B)** is correct. On the tenth day the concentration in blood is about 6.6%.

85. **(A)** is correct. Both the curve for uncrowded and moderately crowded females rise and fall together and therefore are similar.

86. **(D)** is correct. (A) is the innermost major layer of the wall of the eye. (B) is the space that is described by the surrounding iris tissue. (C) is an anterior modification of the choroid layer. (E) is the gelatinous material that fills the posterior cavity of the eye.

87. **(B)** is correct. CALORIES/CUP SKIM MILK − CALORIES/CUP CEREAL = 40 CALORIES (× 2 CUPS) = 80 CALORIES. (A); 180 CALORIES − 150 CALORIES = 30 CALORIES. (C); 40 CALORIES/CUP SKIM MILK. 70 CALORIES/CUP WHOLE MILK. (D); both have the same amount, therefore 15 GRAMS − 15 GRAMS = 0 (E); both have the same amount, therefore 4 GRAMS − 4 GRAMS = 0.

88. **(C)** is correct. 18 GRAMS − 12 GRAMS = 6 GRAMS. (A) 7 GRAMS − 7 GRAMS = 0 (B) is no different than fat for cereal alone. (D) The 15 grams is in the cereal alone. The milk has no starch of relative carbohydrate. (E) Skimmed milk has no fat.

89. **(B)** is correct. 37 GRAMS − 31 GRAMS = 6 GRAMS. (A) There is none. (C) Fats are lipid compounds. (D) Proteins are not carbohydrates. (E) The dietary fibers are clearly in the milk.

90. **(C)** The greater the number of calories consumed is a measure of potential weight gain. (A) The protein content is the same in both types of milk. (B) Cereal with whole milk has 30 calories more per cup than does cereal with skimmed milk. (D) Calories are a measure of available energy. Therefore, whole milk provides more energy than does skimmed milk. (E) The table provides no information about vitamins and therefore no conclusions can be drawn.

91–96. (A) is the hollow ball of cells that results from blastulation. (B) is the process by which the morula becomes a hollow ball of cells with three germ layers. (C) is the cell that results from the fusion of the egg and sperm cells. (D) is the alkaline fluid that protects the sperm cells against the acid environment of the vagina. (E) is the female analog to the penis.
91. **(E)** 92. **(D)** 93. **(A)** 94. **(D)** 95. **(D)** 96. **(C)**

97. **(E)** is correct.

98. **(B)** is correct.

99. **(A)** is correct.

100. **(D)** is correct. (A) If this were true, then the blocking gene should express itself in

the F_2 generation also. (B) If there is no DNA, there is no chromosome. (C) There is no penetrance data given. (E) Genes are not known to switch back and forth.

101. **(A)** is correct. (B) If true, then only the males would express blood type P. (C) As far as is known, each chromosome has only one locus for each specific set of alleles. (D and E) There is no data presented that would support either of these answers.

102. **(B)** is correct. (A and C) are both blood types that express themselves in a 1:1 ratio in the F_1 generation. (D) There is no evidence for this in the data presented. Genes for the same trait are not linked. Linked genes are found at different loci on the same chromosome.

103. **(A)** is correct. (B, C, D, and E) are not reasonable deductions.

104–108. (A) is a starch digesting enzyme produced in both salivary glands and the pancreas. (B) is a protein digesting enzyme produced by the cells of the small intestinal wall. (C) is the protein digesting enzyme produced by the stomach wall. (D) is the mixture of digesting foods and gastric juices. (E) is the portion of bile that assists in the digestion of fats.

104. **(C)** **105. (D)** 106. **(C)** 107. **(A)** 108. **(A)**

109–112. (A) is all populations of species living in an environment. (B) is the biotic community and its physical environment. (C) is a level of stored energy in a typical biotic community. (D) is the terminating community in an ecological succession. (E) is a biome characterized by lack of trees, cold temperatures, and a permafrost the year around.

109. **(D)** 110. **(E)** 111. **(C)** 112. **(A)**

113. **(B)** is correct.

114. **(A)** is correct.

115. **(D)** is correct.

116. **(D)** is correct.

117–120. (A) is the process that helps explain the orderly development of the embryo. (B) is the process that explains cell specialization. (C) is the primitive gut cavity produced during gastrulation. (D) is the process of cell division that leads to a multicellular mass without increase in volume. The cells are smaller than the original zygote. (E) is the process involved in the formation of the central nervous system.

117. **(C)** 118. **(D)** 119. **(A)** 120. **(B)**

Model Essays

Section 2

Because subjectivity enters into the grading of any essay, it is not possible to provide *the* answer to a particular essay question. When you compare your answers with the model answers that follow, keep in mind that there are a number of ways in which they could have been written. Pay attention to the content and the organization of the answer; see whether you have followed a similar style for content and organization.

Question 1

Photosynthesis is an anabolic form of metabolism in green plants. In the presence of primarily sunlight energy, green plants convert energy-poor carbon dioxide into energy-rich glucose. Photosynthesis may be considered in two parts: the *light reactions,* photosystem I and photosystem II, and the *dark reaction,* the Calvin cycle. The green pigment chlorophyll, in conjunction with other photosynthetic pigments, converts radiant energy from the sun into chemical energy of electrons. In photosystem I, in a two-step process, captured light energy boosts electron energy levels to create an electric potential. This potential is channeled along an electron transport system until it is used to phosphorylate ADP to ATP. The ATP is then made available to the Calvin cycle. In photosystem II, captured light energy is also moved along an electron transport system similar to that in photosystem I. In addition to phosphorylating ADP to ATP for the Calvin cycle, photosystem II also splits water (photolysis) making hydrogen available for the reduction of $NADP^+$ to NADPH + H^+. This hydrogen is then made available for the Calvin cycle. A byproduct of photosystem II is molecular oxygen. The enzymes for photosystems I and II are located in the thylokoid membranes of the chloroplast.

The dark reaction, Calvin cycle, is a light independent reaction. It will occur any time that carbon dioxide, NADPH + H^+, and ATP are available. During the dark reaction, carbon dioxide becomes fixed on a 5-carbon sugar, ribulose. This new 6-carbon compound then splits into two molecules of 3-carbon PGAL. ATP then phosphorylates the PGAL and NADPH + H^+ reduces the PGAL. The resulting $NADP^+$ and ADP are then made available for the light reactions. The energy-rich, reduced PGAL is then either converted into 5-carbon sugar to fix more carbon dioxide or ultimately converted into energy-rich glucose. The enzymes for the Calvin cycle are located in the stroma of the chloroplast.

The C_3 plant fixes carbon dioxide through the Calvin cycle while the C_4 plant utilizes another pathway, the Hatch–Slack Pathway. This pathway fixes carbon dioxide into a 4-carbon oxaloaceteate. The leaves of C_4 plants have two different photosynthetic cells; mesophyll cells and surrounding bundle-sheath cells. Oxaloacetate produced in mesophyll cells is changed to asparate or malate which are transferred to bundle-sheath cells where carbon dioxide is removed and refixed by the Calvin cycle. The C_4 plant photosynthesizes two to three times faster than the C_3 plant but at the cost of more ATP and water utilization than the C_3 plant. The C_4 plant grows faster, produces glucose faster per unit area of leaf, and utilizes light energy more efficiently than does the C_3 plant.

Question 2

a. The alimentary tract is essentially a set of tubular structures from mouth to anus. It begins with an opening surrounded by fleshy lips. The inner or oral cavity is composed of the teeth, tongue, hard and soft palates, and cheeks. The teeth are modified for a variety of functions. The incisors are used for chopping and cutting. The canines in humans are poorly developed but are used for grasping in carnivores. The premolars and molars are employed in the grinding of food. The typical tooth is a calcified structure with the crown sticking up above the gum line and the root embedded in the jawbone below the gum line. The main calcified layer of the tooth, dentin, is covered by enamel on the crown and covered with cementum on the root. The center of the tooth is composed of a soft pulp.

The oral cavity leads to the pharynx which serves both the respiratory and digestive systems. The bolus of food created in the oral cavity is moved through the pharynx into the esophagus and on to the stomach. The stomach is a saclike organ that has four distinct histological regions: the cardia, the fundus, the body, and the pylorus. The chyme produced in the stomach is moved from the pylorus into the duodenum, the first segment of the small intestine. The digesting food then moves into the jejunum and on to the ilium, the remaining two segments of the small intestine, in which digestion is completed and absorption occurs. The ilium terminates in the cecum, a blind pouch between the small intestine and the large intestine. Nondigested material moves into the large intestine where water and minerals are absorbed. The remaining compacted mass move temporarily into the rectum and is finally pushed out through the anal canal during defecation.

b. The enzymes responsible for protein digestion arise from three sources; glands in the lining of the stomach, glands in the lining of the small intestine, and glands in the pancreas. The stomach produces gastric juice which contains a number of proteolytic enzymes under the single name pepsin. Pepsin hydrolyzes proteins into polypeptides. The most important function of pepsin is the digestion of collagen, a major constituent of fibrous tissue in meat. The intestinal juice produced by the intestinal wall contains proteolytic enzymes that continue the hydrolysis of polypeptides into amino acids. Another digestive juice is produced in the pancreas and moved into the duodenum through the pancreatic duct. Pancreatic juice contains a variety of endopeptidases (trypsin and chemotrypsin) and exopeptidases (carboxypeptidase and aminopeptidase). The amino acids produced from proteins are then absorbed by active processes in the latter portion of the small intestine.

Question 3

a. Evolution involves change in the frequency of genes in a gene pool over a period of time and results in adaptations to the environment made by a population of organisms.

b. (i) Fossil Record—Fossils are the mineralized remains of plants and animals that lived long ago. The fossil record indicates that the major groups of animals appeared on earth in a sequential manner, the earliest forms being rather simple and the later forms more complex. The evidence for this lies in the fact that the fossils found in progressively newer layers of rocks are those of the increasingly complex forms of animal and plant found along with the older surviving forms. Older layers of rock never contain younger, more complex animal and plant forms.

(ii) Comparative Anatomy—A comparative study of structure in animals in both animals and plants suggests a unity of structure in groups of organisms. Two common examples of unity of structure are flower structure among the angiosperms and skeletal structure among the vertebrates. Organisms can be classified on the basis of anatomical similarity. The unity of plan is based upon the concept of common ancestry. Structures that have arisen through descent from a common ancestor are called homologous structures.

(iii) Comparative Embryology—Groups of organisms share common embryonic developmental stages. Early embryonic stages of all vertebrates bear a striking set of similarities. The human embryo, for example, in its early stages develops gill pouches and develops a tail even though the adult human does not have gills or a tail. During embryonic development the human heart passes successively through 2-, 3-, and 4-chambered stages before developing into the uniquely modified human pattern.

(iv) Vestigial Structures—Occasionally, a human baby is born with a "tail" which has no apparent function in humans. Humans have many other examples including the third eyelid, a leftover of the nictitating membrane in amphibians and reptiles, the vermiform appendix, a leftover from some unknown structure and function, and the pinna muscles, a leftover from mammals that can prick and turn their ears in the direction of a sound. Vestigial structures can be considered the result of genes that are on their way out of a population.

(v) Comparative Biochemistry—Basic biochemical molecules are used in all living organisms, e.g., DNA, RNA, and ATP. Specific proteins can be used to show close or distant relationships among groups of organisms. The closer the relationship, the greater the biochemical similarities.

Question 4

a. Sunlight is the source of energy for all animals and almost all plants. *Producers* (autotrophs) are the organisms that are capable of converting simple, energy-poor inorganic substances into complex, energy-rich organic compounds such as carbohydrates, lipids, proteins, and nucleic acids. Producers often serve as food sources for the *primary consumers* (heterotrophs). Primary consumers are also known as herbivores. Primary consumers often serve as food sources for the *secondary consumers*, (heterotrophs). Secondary consumers are also known as carnivores. It is possible to have tertiary and quarternary consumers in a food chain.

Omnivores are primary, secondary, tertiary, and even quarternary consumers. Death of organisms in any of these levels provide the *decomposers* with food. The decomposers convert complex organic material from nonliving organisms into simple nutrients that go back into the abiotic portion of the ecosystem. These materials are cycled through life again and again.

b. The most probable food web relationships.
The trophic levels for this food web are as follows:

- *Brush vegetation* represent producers, autotrophic organisms
- *Herbivorous insect* and *caribou* represent primary consumers, heterotrophic organisms. (Robin is a possibility here.)
- *Wolf*, *spider*, and *robin* represent secondary consumers, heterotrophic organisms.
- *Hawk* and *robin* represent tertiary consumers, heterotrophic organisms.

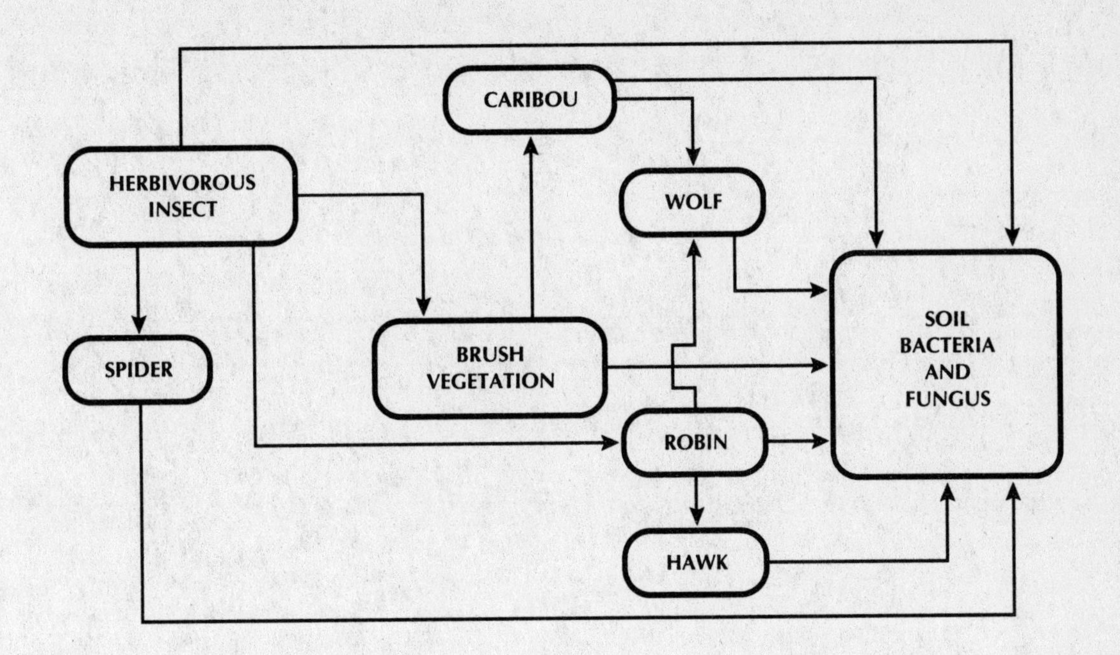
CARIBOU
HERBIVOROUS INSECT
WOLF
SOIL BACTERIA AND FUNGUS
SPIDER
BRUSH VEGETATION
ROBIN
HAWK

PART THREE

IMPORTANT TOPICS FOR REVIEW

To help you prepare for your exam, this section presents capsule reviews of the topics you must know and understand to earn advanced-placement credit in biology. Starting with a brief explanation of biological language, it proceeds to cover those areas of molecular, cellular, population, and organismic biology tested by the AP examination. Each capsule review begins with a list of words to be defined and ends with a miniquiz with explanations for all answers. Because this book is meant to supplement rather than replace your biology textbook, details are kept to a minimum in the topic reviews. The emphasis is on those generalizations and concepts that are essential to understanding the biological sciences and scoring high on the AP examination in biology.

BIOLOGICAL LANGUAGE

The flow of information in science depends on the use of a specific language to convey knowledge, concepts, and principles. Many of the roots, prefixes, and suffixes in scientific terminology have their origins in Latin and Greek. You should learn more commonly used word parts; this will help you to understand previously encountered terms and decipher biological terms with which you are unfamiliar.

See how many of the words in the list below you can define without looking at the explanations that follow.

Biology	Anthropology	Psychology
Embryology	Histology	Cytology

The first thing that you will notice is that all of these terms end in -ology, which means *the study of*. The meaning of the prefix is listed below:

Bio = Life	Anthrop = Man	Psych = Mental Process
Embry = Fetus	Hist = Tissue	Cyt = Cell

There is a danger, however, in always applying a literal interpretation. One example is the term "ecology." The prefix ec means a house. The term "ecology" translates to the study of a habitat or environment. The accepted definition for ecology is a study of relationships between and among organisms and their environment. With practice, you will to develop the valuable skill of defining unfamiliar terms as you encounter them. The following brief list of prefixes and suffixes is provided to help you get a start when dealing with biological terminology.

PREFIXES

a (an) = a negative
ana = return to again
aut = self
chem = chemical
chrom = color
cyt = cell
di = two
ect = outside, without
end = within, inside
ex = out, outside
extr = beyond
glyc = sweet
heter = other, different
hex = six
hom = equal, same
hydr = water
hypo = under, less
hyper = above, beyond
intra = within
is = equal, similar
lys = release, loosen
macr = long, large
meta = next to
micr = small
mon = one
oste = bone
pent = five
phyt = plant
poly = many
prot = first
pseud = false
tetra = four

SUFFIXES

-ase = an enzyme	-itis = inflammation	-mer = part, unit	-sacchar = sugar
-cyte = cell	-kine = movement	-plast = formed	-sis = act of
-gen = origin	-ly = releasing	-phyte = plant	-som = body
			-zoa = animal

As you prepare, you should add to these lists. The list of words below is provided as an exercise. Try to derive the meanings of these words by using the list above. Then compare your meanings with the definitions given.

CHROMOSOME	CYTOGENESIS	CYTOKINESIS
GLYCOLYSIS	HISTOGENESIS	OSTEOCYTE
POLYMER	PROTOZOA	TRISACCHARIDE

chromosome = a body with color (after staining; a hereditary unit within a cell undergoing division)
cytogenesis = the beginning of the cell (how cells form)
cytokinesis = moving the cell (division of the cytoplasm during telophase of mitosis)
glycolysis = releasing sugar (splitting sugar to release energy)
histogenesis = the origin of tissue (tissue formation)
osteocyte = bone cell
polymer = many parts or units
protozoa = first animal
trisaccharide = a three-carbon sugar

MOLECULES AND CELLS

BIOLOGICAL CHEMISTRY

Words To Be Defined

ELEMENT
PERIODIC TABLE
ATOM
HYDROGEN
URANIUM
PROTON
ELECTRON
NEUTRON
ATOMIC NUCLEUS
ORBITAL
ATOMIC NUMBER
ATOMIC WEIGHT
MOLECULE
IONIC BOND
COVALENT BOND
ION
CATION
ANION
IONIZATION
ACID
BASE
BUFFER
INORGANIC COMPOUND
ORGANIC COMPOUND
WATER
CARBON DIOXIDE
CARBOHYDRATE
LIPID
PROTEIN
NUCLEIC ACID
MONOSACCHARIDE
TRIOSE SUGAR
PENTOSE SUGAR
HEXOSE SUGAR
RIBOSE SUGAR
DEOXYRIBOSE SUGAR
GLUCOSE
FRUCTOSE
GALACTOSE
DISACCHARIDE
SUCROSE
MALTOSE
LACTOSE
POLYSACCHARIDE
STARCH
GLYCOGEN
CELLULOSE
NEUTRAL FAT
PHOSPHOLIPID
STEROLS
AMINO ACID
PEPTIDES
PRIMARY STRUCTURE
SECONDARY STRUCTURE
TERTIARY STRUCTURE
QUATERNARY STRUCTURE
NUCLEOTIDES
DNA
RNA
ADENINE
THYMINE
CYTOSINE
GUANINE
TRANSCRIPTION
TRANSLATION

THE ATOM

An *element* is the smallest particle in nature that is indivisible by ordinary chemical means. There are 92 naturally occurring elements that have been identified and organized, by structure and reactivity, into an arrangement called the *periodic table*. The simplest component of an element is the unit called the *atom*. The first and lightest atom is *hydrogen*. The last and heaviest is *uranium*. Other examples of atoms are carbon, oxygen, nitrogen, phosphorus, and sulfur. Atoms are composed of smaller particles known as *protons* (positively charged), *electrons* (negatively charged), and *neutrons* (neutral). The protons and neutrons of a typical atom are clustered together into the *atomic nucleus* while the electrons travel at high speeds around the nucleus. Electrons travel in prescribed three-dimensional paths known as *orbitals* or *energy levels* (Fig. 3-1).

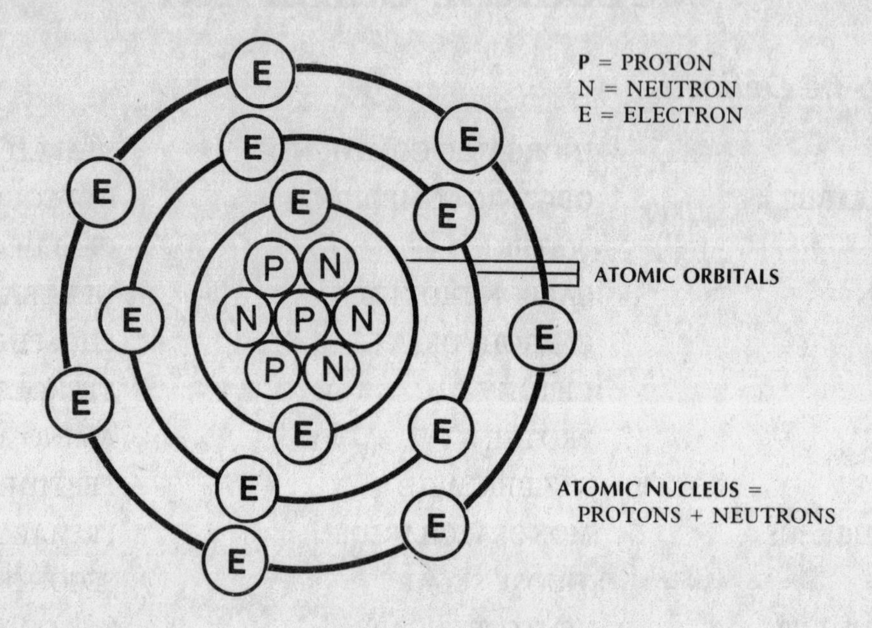

Fig. 3-1 Typical Atomic Structure

Atoms are electrically stable and therefore the number of electrons equals the number of protons. The *atomic number* designates the number of protons that an atom possesses. The *atomic weight (atomic mass)* designates the total of the number of protons and neutrons in the atomic nucleus. The electrons are so small that they effectively do not contribute to the weight of an atom. Given that chlorine has an atomic number of 17 and an atomic weight of 35, you can immediately determine the number of protons (17), number of electrons (17), and number of neutrons (18) present in the atom. The number of neutrons is always determined by subtracting the atomic number from the atomic weight. The chemical properties of an atom are dependent upon the distribution of electrons around the nucleus. When the outermost orbital is stable, the atom will not contribute, accept, or share an electron and therefore will not react chemically.

MOLECULES

Atoms that accept, donate, or share electrons are reactive and participate in the formation of *molecules* (relatively stable aggregations of bonded atoms). When an atom accepts or donates an electron, the bond that results is called an *ionic bond*.

In the case where an atom shares an electron with another atom, the atoms reacting form a *covalent bond*. Ionic bonds are weak-type bonds that break in water, and the ionic bond components then separate. Since ionic substances have either lost or gained an electron in the formation of the bond, their separation leaves each component with either an extra plus or minus charge. Charged atoms are called *ions*. Positively charged ions are called *cations*, while negatively charged ions are called *anions*. Ionic substances such as NaCl separate in water into $Na+$ and $Cl-$. This process is called *ionization*. Ions are necessary in much of biological chemistry.

Two important classes of ionic compound are *acids* (substances that increase the concentration of hydrogen ions in solution) and *bases* (substances that decrease the concentration of hydrogen ions in solution). The acid or base concentration of a solution is measured and recorded as the negative logarithm of the hydrogen ion concentration (pH scale). Neutral solutions have a pH of 7, acid solutions have a pH of less than 7, and basic solutions have a pH of more than 7. The pH of human blood is about 7.4, which means that human blood is normally slightly alkaline. The maintenance of pH at a particular level is very important to normal biological functioning. *Buffers* are substances that either accept hydrogen ions or donate hydrogen ions and therefore control the pH of a solution. Blood has buffers to ensure that it will not get too alkaline or too acidic.

TYPES OF COMPOUNDS

Chemicals in living things are either *inorganic compounds* (simple molecules without carbon chains) or *organic compounds* (complex molecules containing chains of carbon atoms). The three important inorganic substances or groups of substances are *water* (life's solvent), *carbon dioxide* (the basis of life's organic skeleton), and *ions* (the basis of life's physiology). Organic compounds are large, complex molecules composed of carbon-chain skeletons to which are bonded hydrogen atoms, oxygens atoms, nitrogen atoms, etc. The four classes of organic compounds are *carbohydrates* (simple and complex sugars), *lipids* (fats, waxes, oils, and sterols), *proteins* (simple and complex chains of amino acids), and *nucleic acids* (complex chains of nucleotides).

Carbohydrates

These are compounds usually composed of hydrogen, oxygen, and carbon. *Monosaccharides* are simple sugars such as *triose sugars* (composed of three carbons), *pentose sugars* (composed of five carbons), and *hexose sugars* (composed of six carbons). Important sugars of the pentose type are *ribose sugar* and *deoxyribose sugar*. Important sugars of the hexose type are *glucose*, *fructose*, and *galactose*. *Disaccharides* are composed of two sugars. Examples of disaccharides are *sucrose* (glucose + fructose), *maltose* (glucose + glucose), and *lactose* (glucose + galactose). *Polysaccharides* are complex sugars made up of long chains of glucose. The three important polysaccharides are *starch* (principal carbohydrate storage product in higher plants), *glycogen* (principal carbohydrate storage product in animals), and *cellulose* (a principal support material in the walls of plant cells).

Lipids

Forming a wide range of compounds, lipids are principally composed of carbon, hydrogen, and oxygen but may include phosphorus and nitrogen. The best-known lipids are *neutral fats,* composed of a glycerol backbone and three fatty acids. *Phospholipids* are complex lipids that include a phosphate group. *Sterols* are classified with lipids on the basis of their solubilities but differ markedly in structure from the general classes of lipid. Sterols serve as structural elements in membranes of cells, as vitamins, and as hormones.

Proteins

More complex than either carbohydrates or lipids, proteins are made up of chains of *amino acids* or *peptides* (units composed of carbon, hydrogen, oxygen, and nitrogen). There are 20 different kinds of amino acids that go into making up proteins. Proteins are fundamental to structure (globular proteins of membranes) and function (protein hormones and enzymes). All amino acids possess the same fundamental structure (see Fig. 3-2). Each has a central carbon to which is attached a hydrogen, a carboxyl group (–COOH), an amine group (–NH_2), and a side chain (–R). It is the 20 different side chains that make the 20 different amino acids.

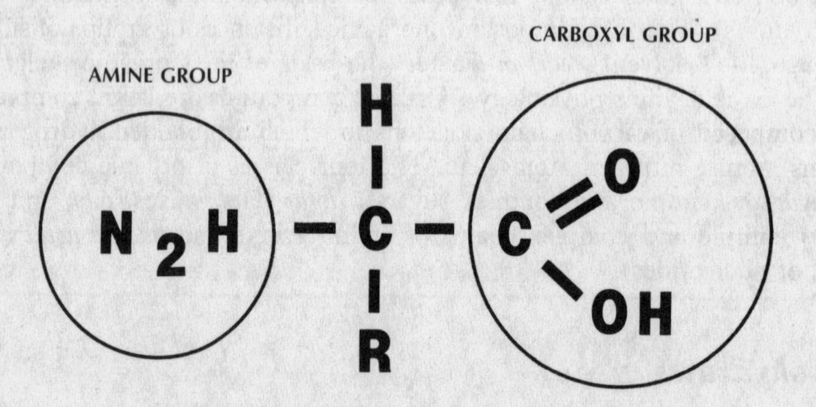

Fig. 3-2 Typical Amino Acid Structure

Primary proteins have *primary structure* (a specific peptide chain composed of a specific number and types of amino acids, e.g., the hormone oxytocin.) Secondary proteins have primary and *secondary structure* (a recurrent alpha pattern: spiral, or a recurrent beta pattern: pleated sheet, e.g., the fibrous protein collagen). Tertiary proteins have primary structure, secondary alpha structure, and *tertiary structure* (a spiral which has been folded into a complicated spherical or globular shape, e.g., the muscle protein myoglobin). Quaternary proteins have *quaternary structure* (a complex of two or more globular proteins, e.g., the protein hemoglobin). There will be a further discussion of proteins in the section on enzymes.

Nucleic Acids

The most complex major class of organic compounds is the *nucleic acids*. They are long chains of units called *nucleotides* (units composed of a 5-carbon sugar, a phosphate group, and an organic nitrogen-containing base). The two general forms of nucleic acid are *deoxyribonucleic acid* (DNA) and *ribonucleic acid* (RNA). DNA is formed by a double strand of nucleotides, each nucleotide containing a deoxyribose (pentose) sugar, a phosphate group, and one of four possible organic bases (either *adenine, thymine, cytosine,* or *guanine*). DNA molecules function as *genes* (nature's units of heredity) and as the blueprint for protein synthesis within the organism. RNA is formed by a single strand of nucleotides, each nucleotide containing a ribose (pentose) sugar, a phosphate group, and one of four possible organic bases (either *adenine, uracil, cytosine,* or *guanine*). RNA functions during the process of protein synthesis. The genetic message for protein synthesis in DNA is passed along as a complement in the synthesis of RNA (*transcription* = DNA makes RNA). The RNA then directs the assembly of a particular protein for which it is encoded (*translation* = RNA makes protein).

MINIQUIZ ON BIOLOGICAL CHEMISTRY

1. A chemical analysis of an organic compound reveals a thymine–uracil ratio of 4:1. This means that

   (A) the specimen contains no carbohydrates or lipids
   (B) 80% of the nucleic acids present are RNA types
   (C) 20% of the nucleic acids present are RNA types
   (D) glucose and galactose are present
   (E) there is a 20% proportion composed of cytosine

2. A quartenary protein is composed of

   (A) an amino-acid sequence only
   (B) a double sequence of nucleotides coupled together with weak hydrogen bonds
   (C) an amino-acid sequence that is coiled and folded into a complicated spherical, or globular shape
   (D) a number of independently folded polypeptide chains
   (E) long-branched chains of monosaccharides

3. If the atomic number of an element is 8 and the atomic weight is 18, then the number of neutrons in the nucleus of the atom is

   (A) 8 (B) 18 (C) 26 (D) 10
   (E) none of these

4. Six molecules of a hexose sugar would contain a total of

   (A) 30 carbon atoms
   (B) 36 carbon atoms
   (C) 18 carbon atoms
   (D) 72 oxygen atoms
   (E) 6 nitrogen atoms

5. A compound that contributes hydrogen ions to a solution is identified as a/an

   (A) acid
   (B) base
   (C) neutral substance
   (D) nucleotide
   (E) atom

6. An example of an organic compound is

   (A) water
   (B) carbon dioxide
   (C) sodium chloride
   (D) glucose
   (E) potassium sulfate

7. Glycogen is an organic compound that is composed of long chains of

   (A) sugars
   (B) amino acids
   (C) nucleotides
   (D) organic bases
   (E) sterols

8. A wide range of organic compounds principally composed of carbon, hydrogen, and oxygen but that may include phosphorus and nitrogen are the

   (A) carbohydrates
   (B) lipids
   (C) proteins
   (D) nucleic acids
   (E) salts

9. The 20 different amino acids found in proteins all possess the same fundamental structure but differ in the

   (A) structure of the carboxyl group in each type
   (B) side chain present in each type
   (C) structure of the amine group in each type
   (D) kind of central atom present in the amino acid
   (E) size of the central carbon

10. One important difference between DNA and RNA is that

    (A) DNA is composed of only a single chain of nucleotides
    (B) RNA is composed of only a single chain of nucleotides
    (C) DNA does not contain a pentose sugar
    (D) DNA is responsible for translation while RNA is responsible for transcription
    (E) only DNA contains uracil

Answers and Explanations for the Miniquiz

1. **(C)** is correct since thymine is unique to DNA while uracil is unique to RNA. A 4:1 ratio means 80% DNA and 20% RNA. The other answers make no sense in relation to the question.

2. **(D)** is the definition of a quartenary protein. (A) describes primary protein. (B) describes the structure of DNA. (C) describes the structure of a tertiary protein. (E) describes the structure of a polysaccharide.

3. **(D)** is correct since the atomic number is the number of protons in the atom, and as atoms are electrically stable, it is also the number of electrons in the atom. The atomic weight is the sum of protons and neutrons and there-

fore the atomic weight–atomic number = neutron number.

4. (**B**) is correct as hexose sugar contains 6 oxygens, 12 hydrogens, and 6 carbons; 6 × 6 = 36.

5. (**A**) This is the definition of an acid. (B) is a hydrogen-ion acceptor. (C and D) are neither acceptors nor donators of hydrogen ions. (E) An ion is a charged atom.

6. (**D**) Organic compounds are large, complex molecules composed of carbon skeletons to which are bound other atoms. (A, B, C, and E) are all simple compounds which lack carbon chains.

7. (**A**) is correct as glycogen is a polysaccharide of glucose. (B) describes proteins. (C) describes nucleic acids. (D and E) are nonsense.

8. (**B**) This is the definition of lipids. (A) is composed of carbons, hydrogens, and oxygens. (C and D) must contain nitrogen. (E) are nitrogen substances.

9. (**B**) as amino acids possess a central carbon to which are attached a carboxyl group, an amine group, a hydrogen, and a specific side chain. (A, C, D, and E) are true of all amino acids.

10. (**B**) is correct since DNA is a double chain of nucleotides in which the sugars are deoxyribose types, and thymine is the unique organic base among the four bases found. RNA is a single chain of nucleotides in which the sugar is ribose, and uracil is the unique organic base among the four bases present. (A, C, D, and E) are all reversals of correct answers.

CELLS

Words To Be Defined

CELL MEMBRANE
PROTOPLASM
NUCLEUS
KARYOPLASM
NUCLEAR MEMBRANE
CYTOPLASM
ORGANELLE
INCLUSION
FLUID MOSAIC MODEL
PASSIVE TRANSPORT
DIFFUSION
OSMOSIS
DIALYSIS
CARRIER MOLECULE
ACTIVE TRANSPORT
CARRIER MOLECULE
ATP
SODIUM–POTASSIUM PUMP
PHAGOCYTOSIS
PINOCYTOSIS

The microscopic examination of thin slices of specimens from living things leads to the inescapable conclusion that most living things are composed of a variety of microscopic units of biological activity that are surrounded by semipermeable membranes and are capable of independent self-reproduction. These units are known as *cells,* some of which exist independently (one-celled animals and plants) while others are highly specialized and dependent on other cells (multicelled animals and plants). Cells vary in size and shape from mycoplasmas measuring 0.1 microns in diameter to ostrich eggs measuring 6 to 8 inches in length (one million times larger than mycoplasmas). It is estimated that the adult human body is composed of some 100 trillion cells.

Cells form the foundation of life, structure, function, and heredity. They are characteristically enclosed in a thin *cell, cytoplasmic,* or *plasma membrane* which is selective with regard to the regulation of what enters or what leaves a cell. The living substance of each cell, *protoplasm,* is organized into two distinct areas: The inner region, the *nucleus,* contains *karyoplasm,* which is bounded by the *nuclear membrane;* the region between the nuclear membrane and the plasma membrane contains the *cytoplasm.* Protoplasm is a complex substance composed of ions (crystalloids) and large molecules such as lipids and proteins (colloids). Many membranous and nonmembranous *organelles* can be identified in both the cytoplasm and karyoplasm. Some of the key organelles can be seen diagrammed in Fig. 3-3. The chart on page 69 summarizes the structure and function of key cell organelles.

CYTOPLASMIC ORGANELLES

ORGANELLE	STRUCTURE	FUNCTION
CENTRIOLE	Short rod composed of 9 sets of 3 fused tubules. Located near the nucleus and Golgi apparatus	Important in the process of spindle formation during mitosis and meiosis
CILIUM	Fingerlike projection containing 9 sets of 3 fused filaments around 2 sets in the center	Involved in movement such as mucus in lungs or eggs in fallopian tubes
CYTOPLASMIC MEMBRANE	Typical bilayer of lipid with proteins embedded in a variety of ways	Serves as a semipermeable barrier between the protoplasm and the external environment
ENDOPLASMIC RETICULUM (ER)	A network of channels and vesicles. Some channels are covered by ribosomes (rough ER) and some have no ribosomes (smooth ER)	Smooth ER can serve as a site of steroid hormone or lipid synthesis or can assist in the breakdown of glycogen to glucose. It can also serve as an internal transport system and cytoplasmic compartmentalizer. Rough ER serves as an internal transport system, cytoplasm compartmentalizer, and site of protein synthesis (on ribosomes)
FLAGELLUM	Long whiplike structure similar to that of a cilium	Involved in movement as in the case of a sperm cell
GOLGI APPARATUS	A network of canals, vacuoles, and vesicles	Serves as a packaging center for cell secretions, and as a site for carbohydrate synthesis
LYSOSOME	Small membrane-bound body	Contains proteolytic enzymes. It acts as an intracellular digestive system
MITOCHONDRION	Double-membrane structure with the outer membrane smooth and the inner membrane thrown into folds (cristae)	Contains enzymes necessary to convert the energy in glucose into ATP for biological work
NUCLEAR MEMBRANE	A double membrane with pores, each membrane similar in structure to the plasma membrane	Serves as a selective barrier between the karyoplasm and cytoplasm
RIBOSOME	Dense body of RNA and protein, usually associated with ER	Site of assembly of amino acids into proteins
CHLOROPLAST	A membrane-bound structure with stacked membrane lamellae inside. These membranes contain chlorophyll	The site of photosynthesis in plant cells
CELL WALL (only plants)	A rigid support structure of cellulose around plant cells	Provides plants with a rigid structure for support

KARYOPLASMIC ORGANELLES

ORGANELLE	STRUCTURE	FUNCTION
NUCLEAR MEMBRANE	A double membrane continuous with the ER of the cytoplasm. Regulatory pores are seen throughout	Contains genetic material necessary for heredity and for the control of protein synthesis
NUCLEAR CHROMATIN	Thread of DNA and protein. Coiled thread forms a chromosome during mitosis or meiosis	DNA encodes genetic information that is inherited or that can be transcribed into RNA for protein synthesis
NUCLEOLUS	Condensations of 5–10% RNA, protein, and a little DNA. No bounding membrane	Function uncertain, but nucleoli may be involved in RNA transcription and in processing of RNA

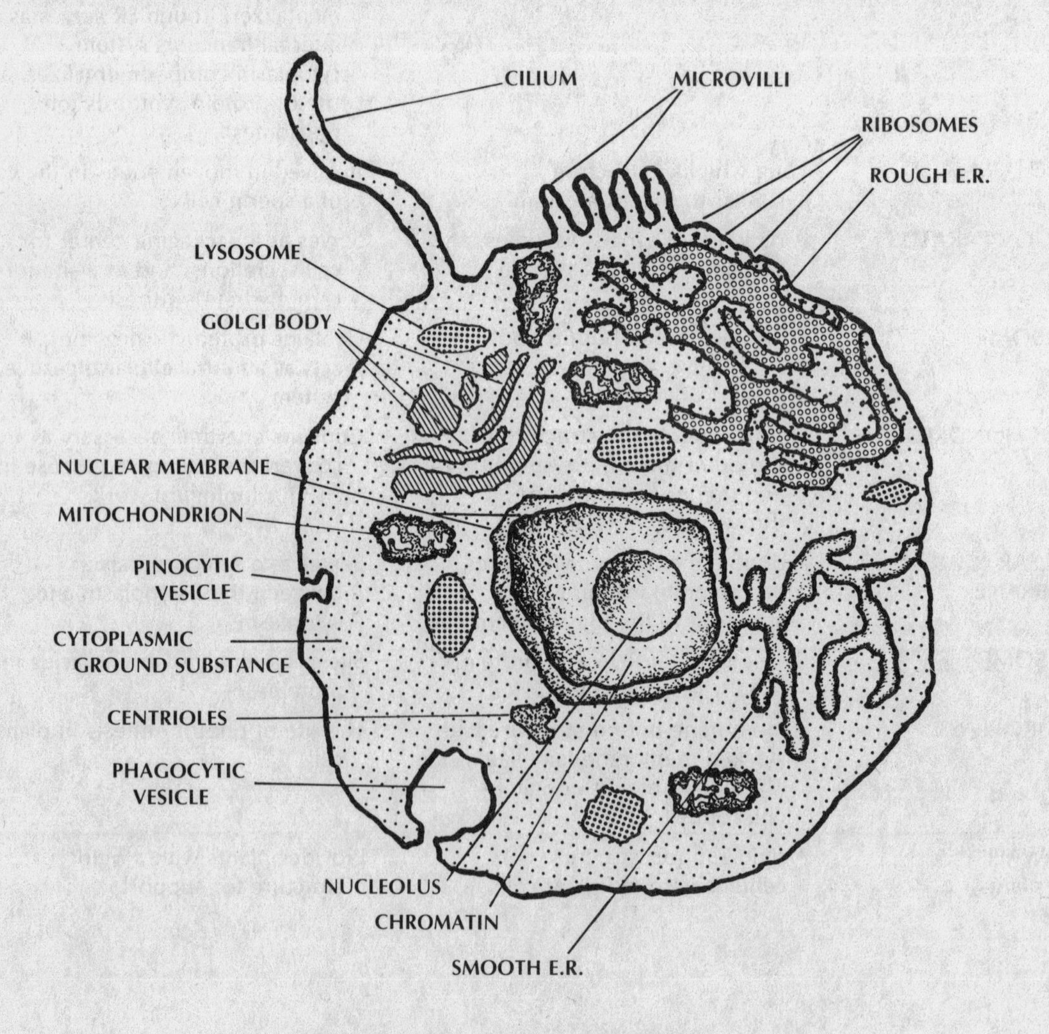

Fig. 3-3 The Typical Animal Cell

In addition to organelles, cells contain *inclusions*. This term covers pigments, stored foods, and crystalline materials.

All membranes—cytoplasmic, nuclear, mitochondrial, endoplasmic, etc.—are composed of lipid, protein, and a little carbohydrate. Electron microscopy had revealed what appears to be a trilaminar structure of all membranes. The original Daveson–Danielle (1941) model proposed a lipid sandwich with a coat of protein on either side. Currently, the Singer–Nicholson *fluid-mosaic* model is the one that is generally accepted. This model envisions membranes as a double-lipid layer "sea" in which protein molecules float like "icebergs" (Fig. 3-4). Data suggest that the hydrophobic ends of proteins stick into the lipid layer and that the hydrophilic ends stick out into the water. It is assumed that the proteins in the lipid "ocean" are in constant motion. The cytoplasmic membrane presents an effective lipid barrier for the cell. Both active and passive transport mechanisms are employed in the movement of substances across the membrane barrier.

Cell survival is dependent on many factors. Membrane transport is perhaps one of the most important. Substances to which membranes are permeable move into and out of cells by means of *passive transport* (a movement along a concentration gradient, but requiring no external energy). When solvent, crystalloid solute, and colloid solute all move along their respective gradients, the phenomenon is known as *diffusion*. In the case where a solvent can move along its gradient but the solutes are restricted by a semipermeable membrane, the phenomenon is known as *osmosis*. In the case where solvent and crystalloid solute move along their respective gradients across a semipermeable membrane but the colloid solute is restricted, the phenomenon is known as *dialysis*.

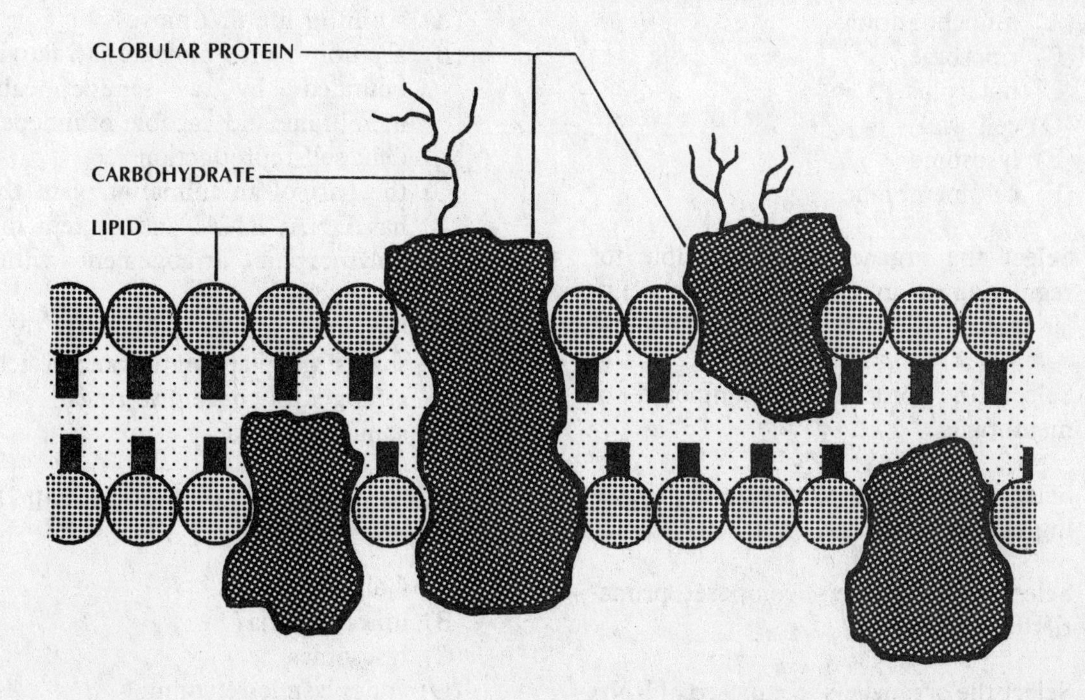

Fig. 3-4 Fluid-mosaic Model of the Cell Membrane

Active transport is a mechanism that requires external energy and can move substances against a concentration gradient. With respect to active transport across a semipermeable membrane, there is usually a *carrier molecule* that forms a complex with the substance to be moved against the gradient. The molecular complex is activated by the presence of external energy in the form of *adenosine triphosphate* (ATP). One example of an active transport mechanism is the *sodium–potassium pump* in the membrane of nerve cells. Sodium ions are present in a relatively high concentration inside nerve cells as compared with the outside. The reverse is true for potassium ions. A stimulus can make the membrane permeable to both sodium and potassium, resulting in a flow of each ion type along its respective gradient across the semipermeable membrane. In order to restore the ion distribution necessary for a response to the next stimulus, the sodium and potassium must be moved against concentration gradients across the semipermeable membrane. It is believed that the membrane contains a carrier molecule that must be energized by ATP prior to moving the sodium and potassium in opposite directions across the membrane.

A different mechanism operates for the movement of large molecules and some fluid components. Saclike invaginations of the cell membrane can form and engulf large molecules or particles (a process called *phagocytosis*) or small volumes of liquid (a process called *pinocytosis*). These invaginations then close and pinch off inside of the cell and bring with them any of the captured materials.

MINIQUIZ ON CELLS

Use the following key for questions 1–5

(A) mitochondrion
(B) ribosome
(C) nucleolus
(D) cell wall
(E) lysosome
(F) cell membrane

1. Select the organelle(s) responsible for regulating the influx and outflux of cellular materials.

2. Select the organelle(s) composed of membranes.

3. Select the organelle(s) capable of producing energy for the cell.

4. Select the organelle(s) composed primarily of cellulose.

5. Select the organelle(s) composed of RNA and protein.

6. The best definition of a cell is

(A) a unit of life that moves
(B) a unit of biological activity bounded by a semipermeable membrane and capable of independent self-reproduction
(C) the part of an animal or plant that has DNA, RNA, and protein in a polymorphic arrangement within the nucleus
(D) the unit of mitochondria, lysosomes, golgi apparatus, and vesicles arranged in concentric rings
(E) something alive

7. Powerful protolytic enzymes will be found in

(A) Golgi apparatus
(B) mitochondria
(C) lysosomes
(D) endoplasmic reticulum
(E) ribosomes

8. According to the best evidence available, cell membranes are composed of

   (A) two layers of protein covered by two layers of lipid
   (B) two layers of lipid in which large proteins are embedded and capable of free movement
   (C) two layers of lipid in which protein molecules are fixed
   (D) three evenly spaced cellulose sheets forming a trilaminar arrangement
   (E) complexes of proteins, lipids, carbohydrates, and nucleic acids

9. The golgi apparatus would be most prominent in cells that

   (A) divide rapidly
   (B) produce ATP
   (C) actively secrete substances
   (D) contain chlorophyll
   (E) are large and inactive

10. The letters ER refer to the

    (A) external ribosomes
    (B) endoplasmic reticulum
    (C) extranuclear regulation
    (D) ergasto regulation
    (E) electron rate of flow

Answers and Explanations for the Miniquiz

1. **(F)** is a lipid barrier that bonds all of the protoplasm and is therefore effective in regulating what enters and what leaves the cell. The cell wall is composed of cellulose but is only found in plant cells. (A, B, and E) are found in the cytoplasm. (C) is found in the karyoplasm.

2. **(A, E, and F)** are all structures composed of the typical membrane. (B) is a nonmembranous condensation of RNA and protein. (C) is a nonmembranous condensation of RNA, protein, and a little DNA. (D) is a nonmembranous structure composed of cellulose.

3. **(A)** is a double-membrane structure, the inner membrane including enzymes necessary in the release of bond energy from glucose. (B) is involved in protein synthesis. (C) is involved in RNA assembly. (D) is a rigid plant structure of the cell. (E) is a membrane structure involved in intracellular digestion.

4. **(D)** Explanation above.

5. **(B)** Explanation above.

6. **(B)** The only objects in nature that can be designated as cells are those that are units of biological activity that are bounded by a semipermeable membrane and are capable of independent self-reproduction.

7. **(C)** Explained in answers 1–3.

8. **(B)** is the fluid-mosaic model of the cell membrane which envisions a bilayer of lipid in which large globular proteins float with their hydrophobic ends inserted into the lipid layer while the hydrophilic ends stick out into the water.

9. **(E)** is composed of membrane channels, vacuoles, and vesicles. It is responsible for packaging of cell secretions. Secretory cells always have well-developed golgi bodies.

10. **(B)** is a common abbreviation for endoplasmic reticulum.

BIOLOGICAL CATALYSTS

Words To Be Defined

CATALYST
ENZYME
SUBSTRATE
ACTIVATION ENERGY
ACTIVE SITE
INDUCED-FIT HYPOTHESIS
PROSTHETIC GROUP
COFACTOR
COENZYME
COMPETITIVE INHIBITION
NON-COMPETITIVE INHIBITION
ALLOSTERIC INHIBITION
NEGATIVE MODULATOR
POSITIVE MODULATOR

A *catalyst* is a substance that speeds up a chemical reaction but itself is unchanged by the reaction. In biological systems, catalysts are special proteins called *enzymes*. Enzymes are very specific in that they (1) usually couple with one set of reactants (*substrates*), and (2) speed up one of the group of reactions for which they are specific. Enzymes play an important role in all chemical reactions within an organism. Most chemical reactions require a reduction in the *activation energy* for reaction initiation. Activation energy is the temporary energy added to a covalent bond allowing the respective atoms to push far enough apart to easily break the bond (Fig. 3-5). The need for activation energy to break covalent bonds is important to the stability of organic molecules, as it prevents them from degrading or breaking down without enzymes. There are times when organic molecules should be degraded and therefore require activation energy. Most chemical reactions require activation energy in the form of heat at temperatures that would be harmful to life processes. In living systems, enzymes speed up chemical reactions at normal physiological temperatures by lowering the activation energy necessary to break covalent bonds. While enzymes do not become involved in the determi-

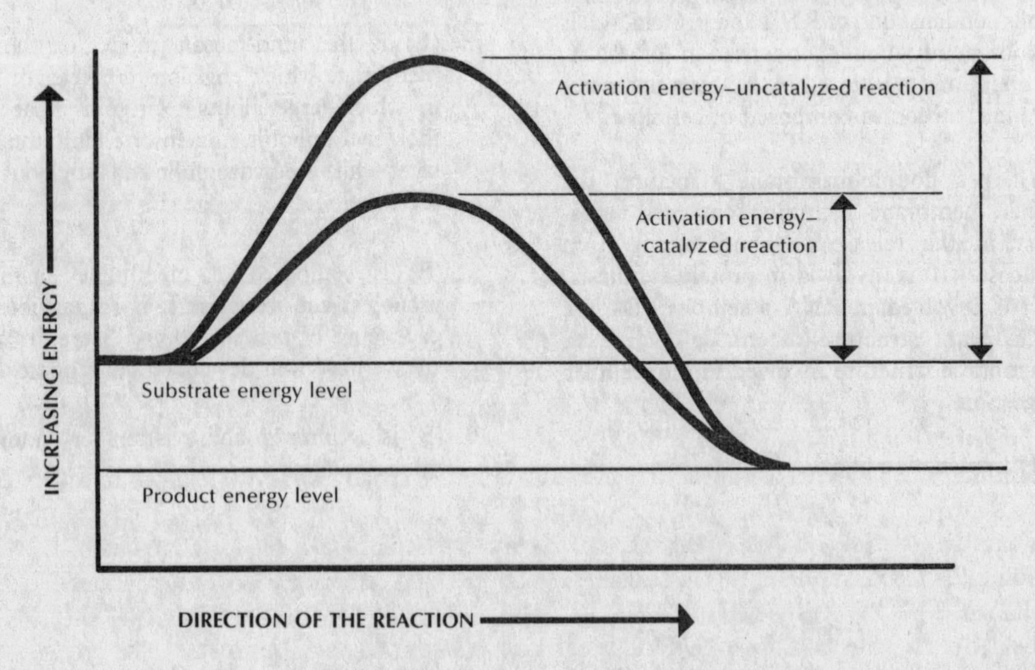

Fig. 3-5 Activation Energy and the Effects of Catalysts

nation of the equilibrium point between two reactants, they are capable of catalyzing forward and reverse reactions.

The tertiary structure of an enzyme (3° globular proteins) confers upon it a unique surface geometry with a specific configuration (the *active site*) that matches a complementary configuration on a specific substrate molecule. This fit is considered similar to a lock-and-key mechanism. The current view of this model is somewhat modified in that there is evidence to suggest that the substrate does not always fit into the active site of the enzyme, and therefore the enzymes must go through a minor conformational change prior to accepting the substrate. This view has been termed the *induced-fit hypothesis*. Many enzymes have a metal-containing *prosthetic group*. Other enzymes lack a prosthetic group and therefore require a *cofactor* that binds temporarily and loosely with the enzyme thereby making it functional. The cofactor *(coenzyme)* may be either a metallic ion or some nonproteinaceous organic substance. As with enzymes, coenzymes are neither used up nor altered by their participation in a chemical reaction.

Several mechanisms have been proposed to explain how enzymes catalyze reactions. Among these are the following hypotheses:

1. The enzyme–substrate complex offers greater opportunity for reaction than that offered when the substrate is free in the fluid medium, because the substrate is being held in an orientation that maximizes molecular collisions for the reaction.
2. The enzyme–substrate complex orients the substrate molecule not only so that its reactive group is oriented to maximize molecule collisions but also so that the fit into the active site increases the reactivity of the substrate.
3. The enzyme–substrate complex induces conformational changes that strain and weaken reactive group bonds making them more susceptible to chemical reaction.

Enzymes are important proteins that control chemical reactions, but they themselves also require control. It is now known that physical factors such as pH, substrate concentration, enzyme concentration, and temperature can influence enzyme activity. There are chemical agents that regulate enzyme activity. Among these chemicals are those that appear to participate in reversible *competitive inhibition*. The inhibitor molecules have shapes that are similar enough to that of a substrate so that they can successfully compete for the active site in the enzyme that is specific for a substrate. The temporary complexing of an enzyme with an inhibitor prevents the enzyme from engaging in a chemical reaction. As enzymes are ordinarily present in small quantities, an inhibitor effectively reduces the general rate of reaction. Other chemicals appear to participate in reversible *noncompetitive inhibition*. These molecules appear to have the ability to attach to a binding site adjacent to an active site that would normally bind a substrate to an enzyme. The bound inhibitor, then, either partially or wholly physically, prevents a substrate from binding to an active site. Another form of noncompetitive inhibition, *allosteric inhibition,* may involve an enzyme that exists in one of two interchangeable conformational forms. One conformational form makes an active site available to substrates while the second conformational form results in the creation of a binding site for a *negative modulator* molecule that binds and stabilizes the enzymes. This second conformational form essentially deactivates the active site by merely changing its shape. There may also be *positive modulators* that stabilize the enzyme molecule in a conformational form that enhances the reactive site binding properties, thereby helping in the acceleration of enzyme activity.

MINIQUIZ ON BIOLOGICAL CATALYSTS

1. Substances that speed up any chemical reaction without themselves being altered in that activity are known as

   (A) enzymes
   (B) catalysts
   (C) substrates
   (D) cofactors
   (E) modulators

2. Activation energy is the energy that is

   (A) required to heat an enzyme to a temperature high enough to induce conformational change
   (B) released when a chemical reaction goes to completion
   (C) produced during the insertion of the positive modulator molecule into the substrate
   (D) needed in the passive transport of the substrate to the active site
   (E) none of these

3. Select the statement that is not true for competitive inhibition.

   (A) It is usually reversible.
   (B) Competitor molecules usually fit into the same active site as the specific substrate that they inhibit.
   (C) The competitor raises the activation energy required for bond breakage.
   (D) The competitor forms a temporary bond with the enzyme.
   (E) The inhibitor generally reduces the rate of reaction for which the enzyme is specific.

4. Enzymes are generally

   (A) primary proteins
   (B) secondary proteins
   (C) tertiary proteins
   (D) quarternary proteins
   (E) none of these

5. Enzyme specificity appears to be a function of the enzyme's

   (A) rate of diffusion
   (B) lipid content
   (C) active site
   (D) activation energy in the enzyme
   (E) none of these

6. A biological catalyst operates on

   (A) its substrate
   (B) its product
   (C) the active sites on carbohydrates
   (D) substrate only at low pH
   (E) coenzymes

7. Enzymes without prosthetic groups usually require

   (A) negative modulators
   (B) positive modulators
   (C) coenzymes
   (D) competitive inhibitors
   (E) none of these

8. Enzymes exposed to temperatures well above accepted physiological limits are most likely to

   (A) speed up the rate of a reaction
   (B) distort their conformational shapes thereby destroying their activities
   (C) convert into carbohydrates that are themselves substrates for that type of enzyme
   (D) increase the available active sites on each molecule
   (E) require two coenzymes rather than one

Answers and Explanations for the Miniquiz

1. **(B)** is the correct statement. The question is expressed in general terms and therefore the answer should be general terms. (A) are biological catalysts. (C) are molecules on which enzymes operate. (D) are required for enzymes that lack prosthetic groups. (E) control enzyme activity.

2. **(E)** is correct. (A) denatures enzymes. (B, C, and D) do not make any sense.

3. **(E)** is not true because competitive inhibitors do not alter the activation energy of an enzyme but they do compete with the substrate for the available active sites. (A, B, C, and D) are all true.

4. **(C)** possess the conformation which produces the active site on the enzyme. (A and B) are too simple to confer the necessary active site conformation on the molecule. (D) is much too large and complex for an enzyme.

5. **(C)** provides a fit for only one kind of substrate. (A, B, and D) make no sense.

6. **(A)** is what the enzyme converts to a product.

7. **(C)** in temporary association with enzymes that lack prosthetic groups allow them to function as enzymes. (A) stabilizes enzymes in such a way as to make the active site unavailable to the substrate and thereby makes them inoperative. (B) stabilizes enzymes in such a way as to make the active site more available to the substrate and thereby enhances the activity of the enzyme.

8. **(B)** Enzymes derive their ability to have active sites by virtue of their three-dimensional conformation. Temperatures above the normal physiological levels usually twist globular proteins out of shape and thereby distort the active sites making them inactive. (A) would be true if not for the fact that high temperatures denature enzymes. (C) is pure nonsense. (D and E) could not be true as the active site is a function of the primary structure of a protein.

ENERGY TRANSFORMATIONS IN LIVING SYSTEMS

Words To Be Defined

ATP
CHEMOSYNTHESIS
PHOTOSYNTHESIS
METABOLISM
ANABOLISM
CELLULAR RESPIRATION
CATABOLISM
PHOTOSYSTEM I
PHOTOSYSTEM II
BENSON–CALVIN CYCLE
PHOTOLYSIS
LIGHT REACTIONS
DARK REACTION
CHLOROPLAST
THYLOKOID
CHLOROPHYLL
CAROTINOID PIGMENT
STROMA
GRANA
GLYCOLYSIS
PYRUVIC ACID
CARRIER MOLECULE
HYDROGEN TRANSPORT SYSTEM
KREBS CITRIC ACID CYCLE
CRISTAE
ANAEROBIC RESPIRATION
AEROBIC RESPIRATION
FERMENTATION
PHOTOPHOSPHORYLATION
OXIDATIVE PHOSPHORYLATION

All work in nature is done with the expenditure of energy. Living things are energy dependent and therefore require mechanisms for energy acquisition, storage, and utilization. Fig. 3-6 summarizes energy's role in life. *Adenosine triphosphate* (ATP) is a molecule that carries energy in high-energy bonds. This energy can be quickly released to bring about biological work; i.e., muscle contraction, active transport, impulse transmission, etc. With the exception of a few *chemosynthetic* bacteria (organisms capable of deriving energy from chemicals in the soil), the source of energy for all other forms of life on earth is *sunlight*. Sunlight energy is captured by life forms that can store the light energy by changing it into chemical bond energy. The energy is stored in the chemical bonds of glucose molecules produced during the process known as *photosynthesis*. Photosynthesis is a form of *metabolism* (the sum total of all of the processes in life) that is called *anabolism* (a constructive process producing complex molecules from simpler ones). When stored energy is required for biological work, the molecules of glucose must be degraded in a process called *cellular respiration*. Cellular respira-

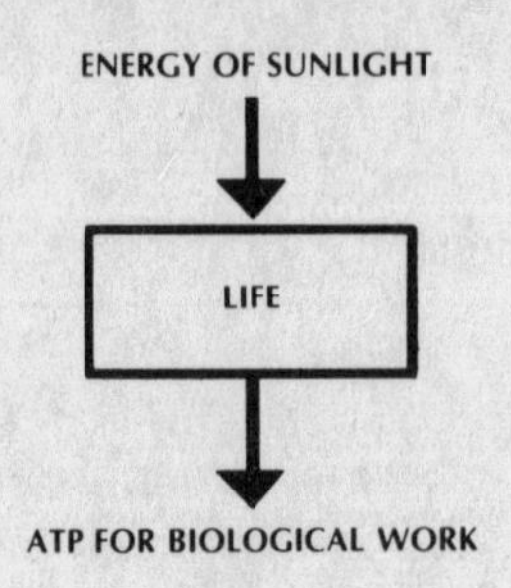

Fig. 3-6

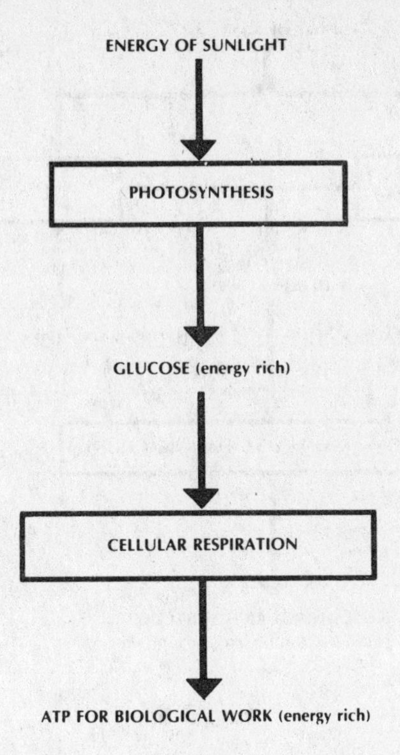

Fig. 3-7

tion is a form of metabolism that is called *catabolism* (a destructive process producing simple molecules from more complex ones). A summary of these processes can be seen in Fig. 3-7.

Photosynthesis employs three specific processes, (see Fig. 3-8). The first two, *photosystem I (cyclic photophosphorylation)* and *photosystem II (noncyclic photophosphorylation),* contain the mechanisms for converting sunlight energy into ATP for the specific work done by the third process, the *Benson–Calvin cycle*. In addition to capturing sunlight energy, photosystem II employs some of the energy in the splitting of water *(photolysis)*, in order to make hydrogen atoms available for use in the Benson–Calvin cycle. The Benson–Calvin cycle uses the hydrogen atoms derived from water and the chemical energy derived from photosystems I and II to fix energy-poor carbon dioxide molecules of the air into energy-rich glucose molecules. As photosystems I and II are light-dependent, they are often referred to as the *light reactions* of photosynthesis. The Benson–Calvin cycle is light-independent and is referred to as the *dark reaction* of photosynthesis. The rate of photosynthesis is dependent upon a number of factors. These factors include temperature, carbon dioxide concentration, and light intensity.

Photosynthesis occurs in cell organelles known as *chloroplasts*. These are membrane-bound structures composed of internally flattened membrane sacs called *thylokoids*. The membranes of the thylokoids contain molecules of *chlorophyll* (the green pigment that captures sunlight energy) and *carotinoid* (orange) *pigments*. Thylokoids are embedded in a colorless protein matrix called *stroma*. A stack of thylokoids is a structure called the *granum*. The enzymes for the light reactions (photosystems I and II) are located in the thylokoid membranes while

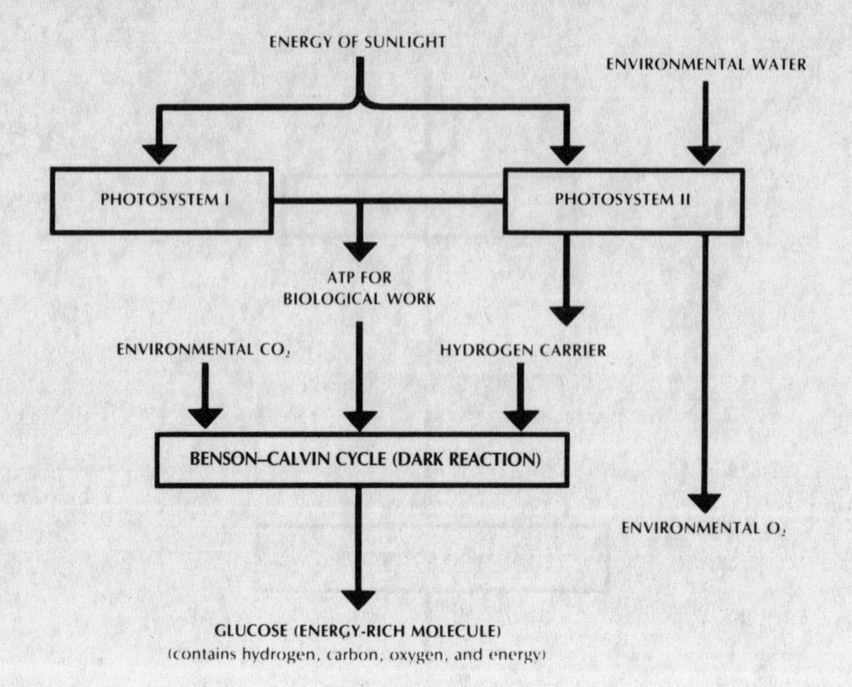

Fig. 3-8

the enzymes for the dark reaction (Benson–Calvin cycle) are located in the fluid stroma. For the specific details of photosynthesis, the reader is directed to any college biology textbook (see Appendix).

Cellular respiration, as with photosynthesis, employs three specific processes concerned with removing the bound energy in glucose in order to make the energy available for some sort of biological work. In the first process, *glycolysis*, glucose is degraded to *pyruvic acid* with the subsequent release of some hydrogen atoms and some of the energy in glucose, (see Fig. 3-9). The hydrogen and energy are picked up by *carrier molecules*, carried to the *hydrogen transport system*, and used in the production of ATP for biological work. At the same time, pyruvic acid is degraded to environmental carbon dioxide in the process called the *Krebs citric acid cycle*. Hydrogen atoms and energy produced in the conversion of pyruvic acid to carbon dioxide, are picked up by additional hydrogen carriers and transported to the hydrogen transport system where they are used in the production of ATP for biological work. Environmental oxygen is necessary to combine with the hydrogen atoms coming off the hydrogen transport system. If not for the formation of water, the hydrogen atoms in water would quickly create a damaging acid environment around these enzyme systems.

Cellular respiration occurs in cell organelles known as mitochondria. A mitochondrion has a smooth outer membrane and an inner membrane with folds *(cristae)* that project into the matrix within the mitochondrion. Some of the enzymes for cellular respiration appear to be free-floating in the compartment between the two mitochondrial membranes. Other enzymes appear to be attached to the cristae, while still others may be free-floating in the matrix into which the cristae project. A total of 36 ATP molecules are produced from the complete breakdown of a single glucose molecule. The hydrogen transport system yields 32

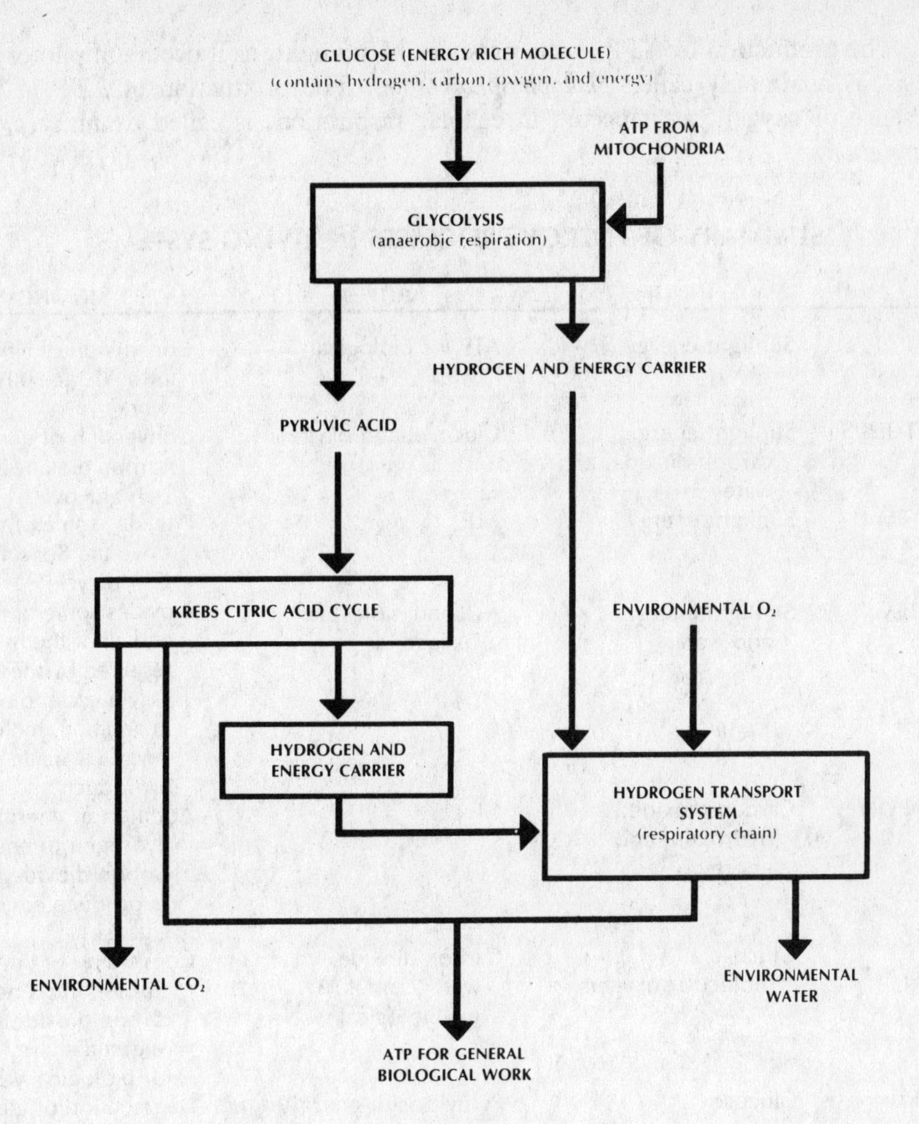

Fig. 3-9

of the 36 ATP molecules. Of the 4 ATP molecules remaining, 2 ATP are produced during glycolysis and 2 ATP are produced directly in the Krebs citric acid cycle. Glycolysis can occur without oxygen and is often referred to as *anaerobic respiration*. The Krebs citric acid cycle and the hydrogen transport system are both dependent upon oxygen and are therefore referred to collectively as *aerobic respiration*. Under anaerobic conditions, glucose is still converted to pyruvic acid, but the Krebs citric acid cycle and electron transport system do not operate. An alternate pathway known as *fermentation* allows pyruvic acid to be converted to either lactic acid or alcohol (dependent upon the type of cell). For the specific details of cellular respiration, the reader is referred to any college biology textbook (see Appendix).

The production of ATP in the presence of sunlight, as it occurs in photosynthesis, is commonly called *photophosphorylation*. The production of ATP in the presence of oxygen, as it occurs in cellular respiration, is called *oxidative phosphorylation*.

SUMMARY OF ENERGY PROCESSES IN LIVING SYSTEMS

PROCESS	INPUT	OUTPUT	SIGNIFICANCE
LIFE	Sunlight energy	ATP for biological work	Conversion of light energy into ATP for biological work
PHOTOSYNTHESIS	Sunlight energy, carbon dioxide, and water	Glucose, and oxygen	Conversion of energy-poor carbon dioxide into energy-rich glucose
PHOTOSYSTEM I	Sunlight energy	ATP	Provides some of the ATP to drive the Benson–Calvin cycle
PHOTOSYSTEM II	Sunlight energy and water	ATP and molecular oxygen	Provides some of the ATP and all of the hydrogen required by the Benson–Calvin cycle to operate. In addition, molecular oxygen is made as a byproduct
BENSON–CALVIN CYCLE	Carbon dioxide, hydrogen, and energy	Glucose	Addition of energy and hydrogen to energy-poor carbon dioxide in order to produce energy-rich glucose
CELLULAR RESPIRATION	Glucose and molecular oxygen	Carbon dioxide, water, and ATP for biological work	Conversion of energy-rich glucose into energy-poor carbon dioxide with the released energy being used for biological work
GLYCOLYSIS (ANAEROBIC RESPIRATION)	Glucose	Pyruvic acid, energy, and hydrogens to be carried away	Degradation of glucose to pyruvic acid with the release of enough energy to produce a net of 2ATP
KREBS CITRIC ACID CYCLE	Pyruvic acid	Carbon dioxide, energy, and hydrogen	Release of remaining energy and hydrogen to be processed in the electron transport system. 2 ATP are produced directly while 32 ATP will be produced in electron transport
ELECTRON TRANSPORT SYSTEM	Energy and hydrogen from glycolysis and Krebs citric acid cycle, and molecular oxygen from the environment	Water and ATP for biological work	Combine released hydrogens with molecular oxygen to produce water and the production of energy-rich ATP

MINIQUIZ ON ENERGY TRANSFORMATIONS IN LIVING SYSTEMS

Use this key in answering questions 1–5

(A) photosynthesis
(B) cellular respiration
(C) Benson–Calvin cycle
(D) anaerobic respiration
(E) photosystem II

1. The process in which energy-poor carbon dioxide is fixed into energy-rich glucose

2. The process in which glucose is degraded to pyruvic acid and 2ATP are produced anaerobically

3. The process during which photolysis occurs

4. The process that is responsible for the conversion of sunlight energy, molecular carbon dioxide, and water into energy-rich glucose and molecular oxygen

5. The process that is light- and water-dependent

6. Select the statement that is *untrue*

   (A) Both plants and animals carry on cellular respiration.
   (B) Animals are incapable of photophosphorylation.
   (C) Glycolysis is a light-dependent process.
   (D) Molecular oxygen of photosynthesis is derived form environmental water.
   (E) Carbon dioxide fixation is a light independent process.

7. Select the statement that is *untrue* for mitochondria.

   (A) Enzymes for the Krebs cycle and electron transport system are found here.
   (B) They contain an inner folded set of cristae.
   (C) They are capable of carrying out oxidative phosphorylations.
   (D) They are capable of carrying out photophosphorylations.
   (E) They require molecular oxygen for their activities.

8. Muscle cells are normally aerobic but fatigue easily when there is too much lactic acid present. Lactic acid in muscle would normally accumulate when there is a/an

   (A) excess of molecular oxygen in a muscle cell
   (B) deficiency of molecular oxygen in a muscle cell
   (C) deficiency of vitamins in cells
   (D) 2° rise in body temperature
   (E) increase in mitochondrial activity

9. The most stable form of chemical energy converted from sunlight energy is found in

   (A) glucose
   (B) ATP
   (C) pyruvic acid
   (D) carbon dioxide
   (E) none of these

10. Thylokoids contain large quantities of

    (A) glucose
    (B) lactic acid
    (C) pyruvic acid
    (D) respiratory enzymes
    (E) none of these

Answers and Explanations for the Miniquiz

1. **(C)** is the correct answer. (A and C) appear to fit the statement and could both be considered good choices, but given that only one answer is acceptable, the one that is specific is the correct choice. (B and D) are out because carbon dioxide fixation has nothing to do with cellular respiration.

2. **(D)** is the correct answer. As with the previous answer (B and D) appear to be valid choices, but again, the specific answer is the better choice. (A, C, and E) have nothing to do with the degradation of glucose.

3. **(E)** Photosystem II supplies the hydrogen and some of the energy needed to drive the Benson–Calvin cycle during the fixation of carbon dioxide. The hydrogens are made available by the splitting of water during noncyclical photophosphorylation.

4. **(A)** is the correct answer. Photosynthesis is the anabolic process that utilizes sunlight energy, water, and energy-poor carbon dioxide in the production of energy-rich glucose.

5. **(E)** Photosystem II is the part of the light reactions that convert sunlight energy into chemical energy of ATP, and split water to make hydrogen available for the Benson–Calvin cycle.

6. **(C)** Glycolysis has nothing to do with photosynthesis. (A, B, D, and E) are all true.

7. **(D)** Photophosphorylation is the production of ATP from the energy available in light. (A, B, C, and E) are all true.

8. **(B)** Lactic acid is produced through an alternate pathway of cellular respiration. This pathway is activated only under anaerobic conditions.

9. **(A)** Glucose is nature's way of "banking" energy for future use. (B and C) are relatively unstable intermediates of biochemical pathways. (D) is a relatively energy-poor compound.

10. **(E)** Thylokoids are chlorophyll-containing membrane structures in chloroplasts. They are the sites where sunlight energy is converted to chemical energy. (A, B, C, and D) are all related to cellular respiration.

REPRODUCTION

Words To Be Defined

EUCARYOTIC CELL
KARYOKINESIS
CYTOKINESIS
DIPLOID CELL
HAPLOID CELL
TRIPLOID CELL
TETRAPLOID CELL
POLYPLOID CELL
MITOSIS
MEIOSIS
CHROMOSOME
HISTONE
CENTROMERE
CHROMATID
CELL CYCLE
INTERPHASE
G_1 STAGE
S STAGE
G_2 STAGE
PROPHASE
METAPHASE
MITOTIC SPINDLE
KINETOCHORE
METAPHASE PLATE
ANAPHASE
POLE
TELOPHASE
MIDDLE LAMELLA
SYNCYTIUM
COENOCYTE
METAPHASE I
TETRAD
CROSSING-OVER
MEIOSIS I
MEIOSIS II
ASEXUAL REPRODUCTION
SEXUAL REPRODUCTION
FERTILIZATION
ZYGOTE

The perpetuation of life is dependent on some sort of cell division. *Eucaryotic cells* (cells that have a distinct nucleus) must first go through a *karyokinesis* (division of the nucleus) before *cytokinesis* (division of the remainder of the cell). Cells normally have nuclei that carry two sets of genetic information (*diploid cells*). Cells that carry only one set of genetic information in their nuclei are called *haploid cells*. Unusual cases exist where some cells may be *triploid* (three sets), *tetraploid* (four sets), or *polyploid* (many sets). There are two types of nuclear division: *mitosis* (diploid nuclei produce identical diploid daughter nuclei), and *meiosis* (diploid nuclei produce haploid daughter nuclei).

The genetic information for life is encoded in molecules of DNA. *Chromosomes* are long, coiled strands of DNA and their associated proteins (*histones*). Before mitosis or meiosis can occur, all of the genetic material must replicate. When chromosome pairs replicate, each resultant pair of *daughter chromosomes* is held together by a single *centromere*. The individual strands are then called *sister chromatids*.

Mitosis

Mitosis is part of the *cell cycle* (that is, the period from the beginning of one mitosis to the beginning of the next mitosis, see Fig. 3–10). The *interphase* is the time from the end of one mitosis to the beginning of the next mitosis. During interphase the cell goes through an extended period of growth activity (G_1), followed by a time when the DNA is replicated (S), and finally, a second, brief period of growth (G_2) prior to the next mitosis.

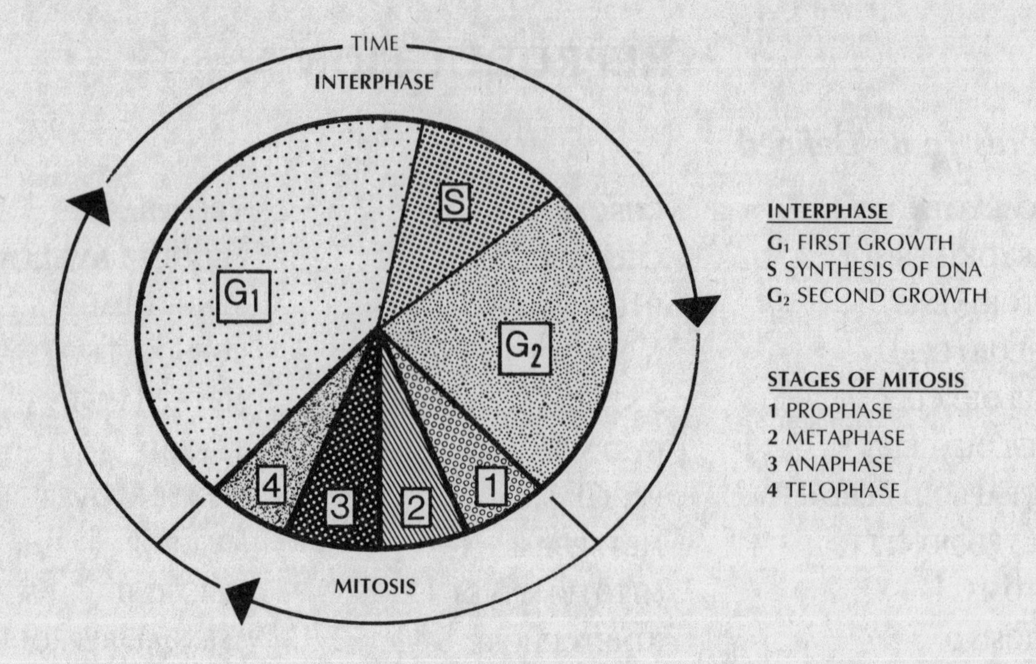

Fig. 3-10 The Cell Cycle

Mitosis occurs in four sequential phases: *Prophase* is the first of these and is characterized by the separation of the centrioles, disappearance of the nuclear membrane and any nucleoli present, and the coiling of the chromatin threads into discrete packages called chromosomes. Each chromosome is composed of two chromatids. *Metaphase*, the second phase, is recognized by the appearance of the *mitotic spindle*, the attachment of the *kinetochore* (a structure in the centromere of each chromatid pair) to a spindle fiber, and the random lining up of the chromosomes along an imaginary *metaphase plate* at the equator of the cell. *Anaphase*, the third phase, is recognized when the centromeres split and members of chromosome pairs begin moving to opposite *poles* (regions to which the centrioles have moved). *Telophase*, the last phase of mitosis, is characterized by cytokinesis; in plant cells this occurs when a *middle lamella* is first established across the middle of the cell, perpendicular to the spindle fibers. This is followed by the development of a cell wall on either side of the lamella and the establishment of a nucleus in each new daughter cell. Cytokinesis in animal cells is characterized by a continuous pinching of the cytoplasmic membrane until two new cells are formed, each with a new nucleus.

In animals and plants there are numerous examples of cells that go through karyokinesis but do not go through cytokinesis. The results are cells that may have from two to several hundred nuclei in each cell. A multinucleate animal cell is called a *syncytium*, while a multinucleate plant cell is called a *coenocyte*.

Meiosis

Meiosis is the form of cell division that requires two divisions of a cell to produce haploid nuclei from a diploid nucleus. Meiosis initially appears to be similar to a typical mitosis. One significant difference occurs during *metaphase I* when chromosome pairs come together in what is known as *tetrad* formation. During this close contact the important event known as *crossing over* occurs. Crossover is a time of mixing of genetic information and provides for variation in gene linkages. During anaphase, tetrad pairs move to opposite poles. Immediately after *meiosis I*, *meiosis II* occurs with no time for the cells to grow or synthesize new DNA. During meiosis II, each nucleus divides once more to form four new haploid cells.

Mitosis forms the basis of growth, and for one-celled organisms, the basis of *asexual reproduction*. Meiosis is stricly part of *sexual reproduction*. Meiosis takes place in either an ovary or a testis; the resultant haploid cells are the egg and the sperm. Fusion of these two haploid cells is the process of *fertilization* and results in the formation of a new diploid cell called the *zygote*. The zygote then begins a number of mitoses leading to the formation of typical multicellular form for a particular species. Animal and plant development will be discussed later.

MINIQUIZ ON REPRODUCTION

1. Crossing-over occurs at the time that the

   (A) tetrads are lined up
   (B) DNA is being synthesized
   (C) zygote begins to go into its first mitosis
   (D) middle lamella has the cell walls forming on either side
   (E) first growth period starts in interphase

2. The stage of mitosis characterized by the appearance of chromosomes, disappearance of the nuclear membrane, and separation of the centioles is

   (A) prophase
   (B) metaphase
   (C) anaphase
   (D) telophase
   (E) interphase

3. The process responsible for equal and identical distribution of genetic information to each new cell is

   (A) fragmentation
   (B) cytoplasmic pinching
   (C) mitosis
   (D) meiosis
   (E) fertilization

4. The statement that is *not* true for meiosis is

   (A) Diploids cells produce haploid cells that are specialized for sexual reproduction.
   (B) Crossover takes place during metaphase I.
   (C) Haploid cells produce diploid cells that become egg or sperm.
   (D) Meiosis is a form of cell division.
   (E) Meiosis occurs only in the reproductive tissues.

5. If the diploid number for the human species is 46, then one polyploid number could be

   (A) 23 (B) 46 (C) 48 (D) 57 (E) 92

Answers and Explanations for the Miniquiz

1. **(A)** is correct since tetrad formation is unique to metaphase I and is the time that crossing over occurs. (B) DNA is synthesized in interphase. (C, D, and E) selections indicate no understanding of meiosis.

2. **(A)** is the correct answer. (B) is characterized by chromosome pairs aligning along the equatorial plate, and the formation of the spindle. (C) is characterized by the separation of the individual members of chromosome pairs. (D) is characterized by either the pinching of the cytoplasm (animal cells), or the formation of the middle lamella and the new cell wall (plant cells), and reestablishment of the new nuclei. (E) is not a part of mitosis.

3. **(C)** is the process that allows a diploid cell to produce two identical daughter cells. (A) is a specific form of reproduction that involves the separation of groups of cells from other groups of cells. (B) is an event that occurs during the telophase of mitosis or meiosis. (D) results in the reduction of the genetic information to one set. (E) is the joining of the egg and sperm nucleus.

4. **(C)** is a reversal of what actually happens. Diploid cells produce haploid cells as a result of meiosis. (A, B, D, and E) are all true.

5. **(E)** is the only choice that is a multiple of the human haploid number. Polyploid cells would have to have multiples of three or more times the haploid number for a species. (A) is the human haploid number. (B) is the human diploid number. (C and D) are aberrant numbers.

GENETICS AND EVOLUTION

MOLECULAR GENETICS

Words To Be Defined

TRANSFORMATION	CONSERVATIVE REPLICATION	MESSENGER RNA
PHOSPHORIC ACID	SEMICONSERVATIVE REPLICATION	TRANSFER RNA
DEOXYRIBOSE SUGAR	DISPERSIVE REPLICATION	RIBOSOMAL RNA
ORGANIC BASE		

Early 20th-century geneticists concluded that the behavior of chromosomes was as would be expected for physical structures that convey genetic information. They later speculated that chromosomes carry the units of heredity, the genes. The early chemical analyses of chromosomes revealed the presence of DNA, RNA, and protein. Because little was known of the nucleic acids, most biochemists believed that the proteins were the hereditary material. Studies of bacterial *transformations* (the processes by which genetic information is transferred from one bacterium to another), began to provide the evidence that it was the nucleic acids and not proteins that encoded genetic information. It was the early 1950s before it was clear that DNA was the genetic material. Careful quantitative measurements of DNA in different body cells eventually demonstrated a consistent quantity from cell to cell, while the quantity of protein in different cells proved to be variable.

In 1953, James Watson and Francis Crick produced the now famous double-helix model of DNA. DNA is now known to be a molecule formed from a double chain of nucleotides held together by weak hydrogen bonds between complementary organic bases (Fig. 3–11). A DNA nucleotide is composed of a *phosphoric acid*, a *deoxyribose sugar*, and an *organic base* (either *adenine*, *guanine*, *cytosine*, or *thymine*). Although there are four organic bases, the complementary arrangements are always adenine to thymine and cytosine to guanine. X-ray diffraction work eventually showed DNA to be a spiraling ladder-like structure with sides formed of alternating units of sugars and phosporic acids. The cross rungs of the ladder are formed by the complementary bases. Each half of the complementary base pair is attached to the sugar of its respective chain.

Once the structure of DNA was understood, it was possible to speculate on how DNA could pass on genetic information. Three key hypotheses were formulated. The first of these was *conservative replication* (The two parent strands of DNA serve as templates for the new double-stranded DNA, thereby leaving the parent molecule intact), the second was *semiconservative replication* (The two parent strands of DNA separate, each acting as a template in the formation of a new strand of DNA); and the third was *dispersive replication* (The two parent strands are broken up into smaller double-stranded segments that were somehow used as templates for

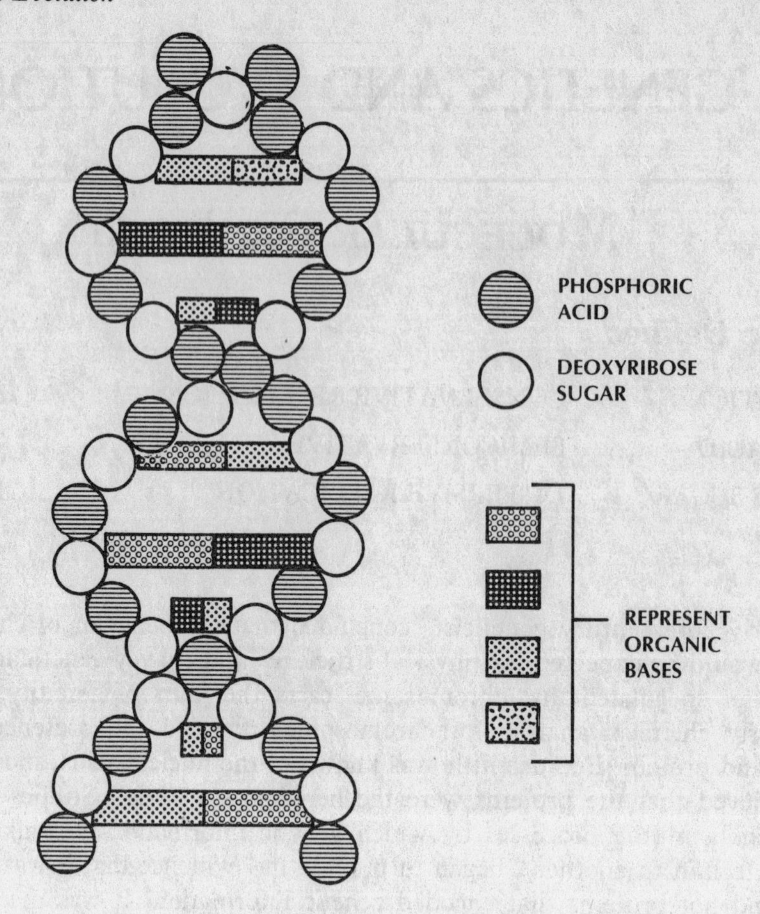

Fig. 3-11 Structure of DNA

the new double-stranded DNA molecule formed). Experimental evidence clearly supports the semiconservative replication mechanism. A variety of enzymes for DNA replication have been discovered and appear to mediate the process.

DNA in an organism determines the structure of the organism's proteins. The DNA first directs the synthesis of specific RNA molecules, which are complementary to one of the two chains in the DNA structure. The RNA then provides the working template for the assembly of amino acids into a polypeptide chain or protein. There are three known types of RNA that are synthesized by DNA. *Messenger* RNA (m-RNA) is a molecule that carries a complementary copy of the DNA code and is capable of determining the amino-acid sequence for a particular protein. *Transfer* RNA (t-RNA) is a molecule that carries a complementary copy of the DNA code and determines which amino acid is to be picked up out of the "amino-acid pool" and inserted into its correct place in the amino-acid sequence being assembled at the ribosome. *Ribosomal* RNA (r-RNA) is a molecule that is a major component of the ribosome. The role of r-RNA in protein synthesis is still not understood.

Discovery of the genetic code, the understanding of its relationship to amino-acid sequence, and the development of techniques for the determination of primary structure (amino-acid sequences) in proteins allow scientists to study evolutionary relationships on a biochemical basis. Comparative protein studies play an important role in the confirmation of evolutionary relationships determined by other comparative studies for example, a study of the fossil record. Of particular use is the identification and study of proteins unique to organisms of a specific line of evolution. The less the variation in an amino-acid sequence, the closer the relationship between two oganisms; the *greater* the variation in an amino-acid sequence, the further apart the relationship between two organisms. One example of a protein unique to the vertebrates is hemoglobin. The hemoglobin of horses and humans differs by about 26 amino acids in each chain, while the hemoglobin of gorillas and humans differs by only one amino acid in each chain. This confirms evidence from the fossil record which shows that gorillas are more closely related to humans than are horses.

GENETIC ENGINEERING

Words To Be Defined

PLASMIDS	TRANSFORMED CELL
RECOMBINANT DNA	CLONED

The manipulation of organismic genes is called *genetic engineering*. The procedure includes: (1) gene isolation and characterization; (2) gene synthesis in the laboratory; (3) coupling of a synthesized gene with regulatory segment of a viral or bacterial gene; and (4) transplantation of the newly constructed gene into a cell where it can undergo replication and directs protein synthesis.

Rings of extrachromosomal DNA, *plasmid,* extracted from bacteria, can be sliced into fragments and coupled with DNA from another source. The plasmid fragments, including the "new" DNA can be reformed into a plasmid. The altered plasmid, *recombinant DNA,* can be introduced into another bacterial cell. This *transformed cell* can now make a protein that it was unable to make before. When cells with new plasmid reproduce cells with the new plasmid, the gene is said to have been *cloned*.

Applications of genetic engineering include: (1) biotic production of protein products that serve as medicines, e.g., insulin and growth hormone; (2) production of protein products not normally synthesized in cells; (3) production of cloned DNA fragments helpful in physician diagnoses of infections and genetic diseases; and (4) alteration of genetic inheritance.

HEREDITY

Words To Be Defined

ALLELE
HOMOZYGOUS
HETEROZYGOUS
DOMINANT ALLELE
RECESSIVE ALLELE
LAW OF DOMINANCE
LAW OF SEGREGATION
MONOHYBRID CROSS
DIHYBRID CROSS
GENOTYPE
PHENOTYPE
LAW OF INDEPENDENT ASSORTMENT
PUNNETT SQUARE

Gregor Mendel's work served as the foundation for early genetic studies. His work was the first experimentally based theory of inheritance. Mendel's success resulted from his accidental choice of pure breeding pea plants, with distinctly contrasting traits which he followed one pair at a time for many generations; large numbers of offspring; his use of statistics in his analyses; and a little bit of luck. Mendel's conclusions can be summarized as follows:

1. Two inherited factors govern each trait studied.
2. One factor comes from each parent.
3. Traits come in several different forms (*alleles*). A condition where there are two of the same alleles for a trait is said to be *homozygous*. A condition where there are two different alleles for a trait is said to be *heterozygous*.
4. In the heterozygous condition, one allele of a trait (the *dominant allele*) may mask the expression of the other allele of a trait (the *recessive allele*). This is known as Mendel's *law of dominance*.
5. Only one of two possible alleles from each parent is passed on to the offspring. This is known as Mendel's *law of segregation*.
6. Mendel performed crosses in which the parents had different alleles for one trait (*monohybrid crosses*) and crosses in which the parents had alleles for two pairs of contrasting traits (*dihybrid crosses*). The results of these crosses led him to conclude that the alleles assorted independently (*law of independent assortment*).

The word "trait" has been replaced by the word "gene". The word "genotype" is used to indicate the genetic makeup of an individual while the word "phenotype" is used to indicate the genetic expression of that individual's genes. If the genotypes of the parents used in a cross are known, then statistical probabilities for offspring bearing certain traits can be determined. The easiest way to do this is to use a *Punnett square*. To construct a Punnett square for a monohybrid cross, the genes present in the gametes of one parent are written along the x-axis of the square, and the genes present in the gametes of the other parent are written along the y-axis of the square (see below). It makes no difference which genes are written along which axis.

Let us look at one example of a monohybrid cross. In pea plants, pure breeding plants for red flowers (RR) can be crossed with pure breeding plants for white flowers (rr). Upper-case letters are used to indicate dominant genes, while lower-case letters are used to indicate recessive genes. Look at the problem as it is worked out below.

(Parents) **RR** × **rr**

The Punnett square for this cross:

	R	**R**
r	**Rr**	**Rr**
r	**Rr**	**Rr**

All of the offspring resulting from this cross will be genotypically heterozygous (100%) but will phenotypically produce plants that produce red flowers (100%). Let us cross the members of the first generation and see what happens.

(Parents) **Rr** × **Rr**

The Punnett square for this cross:

	R	**r**
R	**RR**	**Rr**
r	**Rr**	**rr**

The results of this cross indicate that phenotypically, 3 out of 4 offspring (75%) will be plants that produce red flowers, while 1 out of 4 offspring (25%) will be plants that produce white flowers. Genotypically, 1 out of 4 offspring will be homozygous dominant (25%); 1 out of 4 offspring will be homozygous recessive (25%); and 2 out of 4 offspring will be heterozygous (50%). In summary, the phenotypic ratio is 3:1 while the genotypic ratio is 1:2:1.

Dihybrid crosses involve two independently assorting sets of traits and therefore require a Punnett square that is 4 by 4 (16 boxes). Let us work with the following example.

In pea plants, some plants produce pea coats that are smooth (S) and some plants produce pea coats that are wrinkled (s). An independent trait is for pea coat color. Some plants produce pea coats that are yellow (Y) and some produce pea coats that are green (y). A cross between a plant that is pure bred for smooth, yellow pea coats and a plant pure bred for wrinkled, green pea coats will produce the following F_1 generation:

(Parents) **SSYY × ssyy**

The Punnett square for this cross:

	SY	SY	SY	SY
sy	SsYy	SsYy	SsYy	SsYy
sy	SsYy	SsYy	SsYy	SsYy
sy	SsYy	SsYy	SsYy	SsYy
sy	SsYy	SsYy	SsYy	SsYy

All of the offspring resulting from this cross are genotypically 100% heterozygous for both the pea-coat surface and pea-coat color. Phenotypically, 100% of the plants produce peas with smooth, green coats. Now let us cross members of the first generation and determine genotypic ratios and phenotypic ratios.

:SsYy × SsYy

	SY	Sy	sY	sy
SY	SSYY	SSYy	SsYY	SsYy
Sy	SSYy	SSyy	SsYy	Ssyy
sY	SsYY	SsYy	ssYY	ssYy
sy	SsYy	Ssyy	ssYy	ssyy

The results of this cross indicate that the phenotypic ratio will be 9:3:3:1. Ideally, 9 out of 16 plants will produce peas that have smooth yellow coats, 3 out of 16 plants will produce peas that have smooth green coats, 3 out of 16 plants will produce peas that have wrinkled yellow coats, and 1 out of 16 plants will produce peas that have wrinkled green coats. Genotypically, the ratio is more complex.

	Pea-Coat Surface	Pea-Coat Color
1 out of 16	Homozygous dominant	Homozygous dominant
2 out of 16	Homozygous dominant	Heterozygous
1 out of 16	Homozygous dominant	Homozygous recessive
2 out of 16	Heterozygous	Homozygous dominant
4 out of 16	Heterozygous	Heterozygous
2 out of 16	Heterozygous	Homozygous recessive
1 out of 16	Homozygous recessive	Homozygous dominant
2 out of 16	Homozygous recessive	Heterozygous
1 out of 16	Homozygous recessive	Homozygous recessive

There are some monohybrid and dihybrid cross problems at the end of this section, and you should find additional problems in your textbook. Solve them all, for the more practice you get, the easier it will be for you to work genetics problems.

Modern Heredity

Words To Be Defined

HOMOLOGOUS
LOCUS
LINKED GENE
INCOMPLETE DOMINANCE
LETHAL ALLELE
CARRIER
METABOLIC DISEASE
MULTIPLE ALLELE
CODOMINANCE
POLYGENIC TRAIT
SEX CHROMOSOME
AUTOSOME
SEX-LINKED GENE
SEX-INFLUENCED GENE
NONDISJUNCTION
TRANSLOCATION
DOWN'S SYNDROME
MUTATION

Modern genetic principles are somewhat different from the principles set out by Mendel. The physical evidence that confirmed what Mendel described from indirect observations came with the discovery of chromosomes and the description of the process of meiosis. We now know that chromosomes come in *homologous* pairs (same length, same centromere position, and same gene loci). A *locus* is a position on a chromosome where a gene for a particular trait is located. A locus is occupied by only one allele for a trait. The most recent evidence supports the idea that possibly hundreds or even thousands of genes are arranged in a linear order on each chromosome. Any genes that occupy the same chromosome but different loci are said to be *linked genes.* Since each chromosome moves as a unit during meiosis, the two linked genes must move together rather than assort independently. Specific linkages can be altered somewhat by crossing over tetrad formation in metaphase I of meiosis.

Many allelic pairs of genes are now known to exhibit *incomplete dominance,* a condition in which the heterozygote resembles neither the homozygous dominant phenotype nor the homozygous recessive phenotype. An example of incomplete dominance can be found in the "Four O'clock" plant. The color of the flowers is determined by alleles that exhibit incomplete dominance. When a pure trait red flower plant is crossed with a pure trait white flower plant, the result is a generation of plants that *all* produce pink flowers.

W = White flower
W′ = Red flower
WW′ = Pink flower

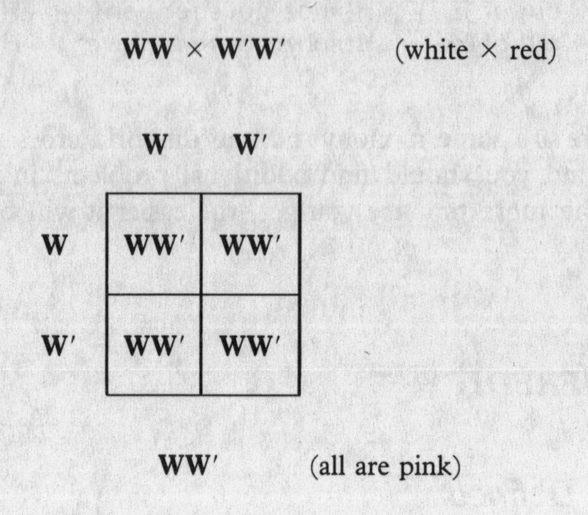

WW × W′W′ (white × red)
W
W′
W
WW′
WW′
W′
WW′
WW′
WW′ (all are pink)

If we now cross two pink flower plants, we obtain a generation that exhibits all three phenotypes.

WW′ × WW′ (pink)

	W	**W′**
W	**WW**	**WW′**
W′	**WW′**	**W′W′**

25% white: 50% pink: 25% red

Some alleles are known to code for no proteins, while others are known to code for inactive proteins. As these alleles are usually recessive, in the heterozygous condition they will not be "seen"; it is the dominant allele that expresses itself. In the homozygous recessive condition, these alleles will have effects dependent upon the importance of the protein that they either do not produce or produce in an inactive state. If the protein is essential to biological functioning, the organism usually dies. These would then be *lethal alleles*. Every human carries an estimated 25–35 lethal alleles. A person heterozygous for a particular trait is said to be a *carrier*. The best-known example of a human lethal gene is the one for sickle-cell anemia.

There are dominant lethal alleles in nature but recessive lethal alleles are more common. Dominant lethal alleles are usually eliminated automatically as there is no way to "hide" the allele. Some recessive alleles code for abnormal proteins and are not lethal because the protein for which they are miscoded is not part of an essential biological pathway. These alleles do not cause the death of an individual but may result in a *metabolic disease*. Two of the classic metabolic disorders related to a genetic defect are phenylketonuria (PKU) and albinism (lack of pigmentation).

The phenomenon of *multiple alleles* involves a number of alleles for a particular gene locus on a chromosome. One of the best-known examples is the human ABO blood group system. Any one of these three alleles can occupy the same locus. The A and B alleles are *codominant*, meaning that they both express themselves when they are present together. Both A and B are dominant to the O allele. This is why these three alleles are responsible for four blood phenotypes in humans. AA and AO are expressed as blood type A. BB and BO are expressed as blood type B. AB express themselves as blood type AB. OO expresses itself as blood type O. Multiple alleles occupy a single locus but there are also *multiple loci* (*polygenic traits*). Alleles at several different loci are all required to influence the expression of a particular trait. The loci may be on either homologous or nonhomologous chromosomes. Skin color in human beings is perhaps the best-known example of polygenic traits.

Up until now we have been dealing with genes and loci. Sex determination in higher animals is the result of arrangements of a single pair of sex chromosomes. One sex is heterozygous for a complete set of *sex chromosomes* and the other sex is homozygous for a complete set of sex chromosomes. In mammals, the male is heterozygous, having one X chromosome and one Y chromosome while the female is homozygous, having two X chromosomes. The nonsex chromosomes are termed *autosomes*. Any defective gene associated with either the X or Y chromosome would be *sex-linked*. There are a number of sex-linked diseases found in humans. Among these are hemophilia and red–green colorblindness. *Sex-influenced genes* depend upon the sex of an individual for their expressions. In humans, the "male pattern baldness" gene is located on an autosome. It acts as a dominant gene for those men who have it and as a recessive gene for those women who have it.

Occasionally, the phenomena of *nondisjunction* and *translocation* occur. Nondisjunction is the lack of separation of homologous chromosomes during meiosis. The result is a gamete that has one extra chromosome or a gamete that lacks one chromosome. One of the best-known examples of nondisjunction in humans is the extra chromosome 21 found in individuals with *Down's syndrome*. Translocation occurs when part or all of one chromosome attaches to another chromosome.

One important phenomenon in genetics is *mutation*, an inheritable change in the genetic material. Mutations are unpredictable and in humans have been estimated to be from 1 in 10,000 to 1 in 1 million. At the molecular level a mutation is the deletion, addition, or change in an organic base in the DNA sequence.

MINIQUIZ ON GENETICS AND EVOLUTION

1. The conclusion that *cannot* be attributed to Gregor Mendel is that

   (A) two inherited factors govern each trait
   (B) every human carries a number of lethal alleles
   (C) one factor comes from each parent
   (D) in the heterozygous condition, one allele for a trait may mask the expression of the other allele for the trait
   (E) only one of two possible alleles from a parent can be passed on to each offspring

2. The expression of a gene in an individual is termed the

   (A) genotype of the individual
   (B) incomplete dominance in the individual
   (C) linked gene
   (D) locus
   (E) phenotype of the individual

3. Only one of two possible alleles from each parent is passed on to each offspring is a statement of the

   (A) law of dominance
   (B) law of segregation
   (C) law of independent assortment
   (D) law of incomplete dominance
   (E) law of nondisjunction

4. A man with blood type A, whose mother had blood type O, marries a woman with blood type B, whose father was blood type O. The chances of this couple's having a child with blood type O are

   (A) 1:4 (B) 1:5 (C) 3:4 (D) 1:2 (E) 9:3:3:1

5. A typical phenotypic ratio for a dihybrid cross is

   (A) 9:1 (B) 3:4 (C) 9:3:3:1 (D) 1:2:1:2:1 (E) 6:3:3:6

6. Metabolic diseases are usually the result of a/an

   (A) lethal allele that is recessive
   (B) lethal allele that is dominant
   (C) nondisjunction of sex genes
   (D) incomplete dominance as seen in the "Four O'Clock" plant
   (E) codominant alleles

7. The position of a gene on a chromosome is known as the gene's

   (A) centromere
   (B) phenotype
   (C) genotype
   (D) expression
   (E) none of these

8. In the controversy over whether protein or DNA was the genetic material, one of the earliest pieces of evidence in favor of DNA was the

   (A) result of bacterial transformation studies
   (B) discovery that certain viruses were composed of RNA and protein
   (C) report by Watson and Crick that DNA had a helical structure
   (D) realization that hemoglobin in humans and gorillas differed by only a single amino acid
   (E) enunciation of the cell theory by Schleiden and Schwann

9. The best evidence supports the hypothesis that DNA replication is by means of

   (A) protein synthesis
   (B) conservative replication
   (C) semiconservative replication

(D) dispersive replication
(E) photographic replication

Use the following key in selecting answers for questions 10–15

(A) messenger RNA
(B) transfer RNA
(C) structural RNA
(D) DNA
(E) protein

10. The product of transcription and translation is _____

11. The molecule that is the repository of the genetic code is the _____

12. The molecule in which mutation usually occurs is _____

13. The molecule that contributes to the phenotype of a gene is _____

14. The molecule that directs the actual assembly of amino acids is _____

15. The molecule that is an integral part of the ribosome is _____

16. Phenylthiocarbamide (PTC) is a bitter chemical that can be tasted only by those who have the dominant allele for the taste sensation caused by the chemical. Those who are homozygous for the recessive allele are "nontasters". When a nontaster and a taster have a number of children, what proportion of their children would you expect to be nontasters? If the taster was homozygous, what proportion of the couple's children would you expect to be nontasters?

17. A couple composed of a taster and a nontaster have four children, all of whom are tasters. What are the genotypes of the parents? What are the genotypes of the four children?

18. The gene for brown eyes is dominant to the gene for blue eyes. A blue-eyed man marries a brown-eyed woman whose father was blue eyed. What would be the the expected genotypes and phenotypes of their children? What would be expected proportions?

19. In the fruit fly *Drosophila*, the allele for normal body (B) is dominant over the allele for hairy body (b) and the allele for normal leg (L) is dominant over the allele for short leg (l). Set up the Punnett square for each of the following crosses.

Cross no. 1 **BBll × BbLl**
Cross no. 2 **BbLl × BBLl**

20. In pea plants, the allele for tall (T) is dominant over short (t), and the allele for yellow pea coat (Y) is dominant over the allele for green pea coat (y). In a plant that is heterozygous for both sets of alleles, what are the chances of a gamete carrying a

(a) T allele? (b) t allele and a Y allele?
(c) T allele and a Y allele?

21. A cell transformed by recombinant DNA, reproduces and makes many copies of a "foreign" gene. The gene is then considered

(A) cloned
(B) probed
(C) mutated
(D) (A and B)
(E) none of the above

22. A plasmid carrying a foreign gene is referred to as

(A) a retrograde cell
(B) a codon
(C) recombinant DNA
(D) (A and B)
(E) none of the above

Answers and Explanations for the Miniquiz

1. **(B)** Mendel never knew anything about lethal genes. (A, C, D, and E) are conclusions deduced by Mendel from his pea studies.

2. **(E)** This is the definition of the term "phenotype." (A) identifies the two alleles present in an individual. (B) is the interaction of two alleles that have an expression different from the phenotype of either gene alone. (C) relates to genes at different loci on the same chromosome. (D) is the location of a gene on a chromosome.

3. **(B)** is correct. (A) accounts for the simple behavior of alleles. (C) accounts for the behavior of unrelated allelic pairs in crosses involving several traits. (D and E) are not laws.

4. **(A)** is correct based on the information available. The man and woman are obviously genotype AO and therefore have a 1:4 chance of having an OO child. (C and D) are real monohybrid ratios but do not apply in this case. (B) is not a reasonable choice as monohybrid crosses have only four possible combinations of maternal and paternal alleles. (E) is a typical phenotypic ratio for a dihybrid cross.

5. **(C)** is correct since typical phenotypic ratio for a dihybrid cross must add up to 16.

6. **(A)** is correct since lethal alleles are usually recessive.

7. **(E)** is correct because the answer required is "locus." (A) is the structure that holds the chromosome pairs together in mitosis and meiosis, and is also the structure to which the spindle attaches. (B) is the expression of a gene. (C) identifies the two alleles present in an individual. (D) is meaningless.

8. **(A)** is correct. (B, C, and D) are related to molecular genetics well after the establishment of DNA as the genetic material. (E) has nothing to do with genetics.

9. **(C)** is the best hypothesis in light of present research. (A) is unrelated to the question. (B and D) are not likely in light of current research. (E) is nonsense.

10. **(E)** is correct. Transcription occurs when DNA makes RNA and translation is when RNA directs the assembly of a protein.

11. **(D)** is the molecule in which the organic base sequence, taken in linear triplets, codes for the assembly of specific proteins.

12. **(D)** is correct since a mutation is a single change in the DNA base sequence.

13. **(E)** is correct as what one sees is the result of protein synthesis.

14. **(A)** is correct. (B) is responsible for bringing a specific amino acid to the ribosome but it is the m-RNA that makes sure that the amino acid is correctly placed into the polypeptide chain. (C) is a component of the ribosome but its role in protein synthesis is not understood.

15. **(C)** is correct.

16. This problem is a monohybrid cross. Let us use T = tasters and t = nontasters. From the information available we can set up the following:

    tt × T?

    Note that we do not have all of the genotypic information on the taster and so we have to use a question mark. If the taster was homozygous, then we would expect none of the children to be nontasters (tt). All of the children would be carriers (Tt). If the taster was heterozygous (Tt), then there would be a 50% chance of having a nontaster child.

17. In this situation, we again do not have the complete genotypic information for one parent. The nontaster parent is clearly (tt) but the taster parent must be presumed to be T? even if the couple has four taster children. All that we can tell about each of the four children is that he or she is T?.

18. In this monohybrid cross, we can write out the full genotypes of the couple because all of the information is given. The woman is brown eyed but had one blue-eyed parent and must be Bb while the man must be bb. They would have a 50% chance of having a blue-eyed child (bb) and a 50% chance of having a brown-eyed child who would be heterozygous (Bb).

19.

BBll × BbLl

	Bl	**Bl**	**Bl**	**Bl**
BL	**BBLl**	**BBLl**	**BBLl**	**BBLl**
Bl	**BBll**	**BBll**	**BBll**	**BBll**
bL	**BbLl**	**BbLl**	**BbLl**	**BbLl**
bl	**Bbll**	**Bbll**	**Bbll**	**Bbll**

BbLl × BBLl

	BL	**Bl**	**bL**	**bl**
BL	**BBLL**	**BBLl**	**BbLL**	**BbLl**
Bl	**BBLl**	**BBll**	**BbLl**	**Bbll**
BL	**BBLL**	**BBLl**	**BbLL**	**BbLl**
Bl	**BBLl**	**BBll**	**BbLl**	**Bbll**

20. A heterozygous plant's genotype is TtYy. As only one of each allelic pair will be found in the gamete, we can write the possible gametes:

TY Ty ty ty

(a) would have a 50% chance of being present.
(b) would have a 25% chance of being present.
(c) would have a 25% chance of being present.

21. **(A)** is correct. (B) refers to a cell exposed to a DNA probe. (C) refers to a single base change in the genetic code on a gene.

22. **(C)** is correct. (A) is nonsense. (B) is a three organic base sequence in the genetic code.

EVOLUTION

Words To Be Defined

SPONTANEOUS GENERATION
PROTEINOID
COACERVATE DROPLET
MICROSPHERE
CELL
HETEROTROPH
AUTOTROPH
EUKARYOTIC CELL
EVOLUTION
LAMARCKISM
DARWINISM
ARTIFICIAL BREEDING
GENE POOL
NATURAL SELECTION
STABILIZING SELECTION
DISRUPTIVE SELECTION
DIRECTIONAL SELECTION
INDUSTRIAL MELANISM
EVOLUTIONARY ADAPTATION
ENVIRONMENTAL ADAPTATION
PHYSIOLOGICAL ADAPTATION
MACROEVOLUTION
MICROEVOLUTION
DIVERGENT EVOLUTION
CONVERGENT EVOLUTION
PARALLEL EVOLUTION
COEVOLUTION
POPULATION GENETICS
POPULATION
HARDY-WEINBERG PRINCIPLE

THE ORIGIN OF LIFE

Prior to the 19th century, it was commonly believed that many forms of life sprang from nonliving materials, or were the subject of *spontaneous generation,* when there were favorable conditions: maggots from rotting meat, frogs from mud, fish from leaves that fell into the water, etc. Starting with the experiments of Francesco Redi and followed by the work of many others, such as Leeuwenhoek and Pasteur, it became clear that life comes from preexisting life and that it gradually diversifies. Spontaneous generation is, however, the basis for the origin of life on earth.

Many scientists today believe that the spontaneous generation of life on earth was inevitable given the conditions of the primitive earth. The early earth is believed to have had a reducing atmosphere rather than the oxidizing one now encircling it. Today's atmosphere precludes any spontaneous generation. The primitive atmosphere, the result of violent volcanic activity, was most likely composed of molecular hydrogen (H_2), methane (CH_4), ammonia (NH_3), and water vapor (H_2O). There is geological and astronomical evidence to support this assumption. The energy necessary to convert inorganic substances into organic materials most likely came from lightning, heat from within the earth, and ionizing radiations from the sun. The one final requirement for the origin of life on earth was time. The Russian scientist Alexander Oparin (1922) and the British scientist Joseph B.S. Haldane (1929) independently speculated that life arose by spontaneous generation from simple inorganic molecules of the primitive earth. It was not until 1953 that Stanley Miller was able to provide the experimental evidence to support Oparin's and Haldane's hypothesis. Miller experimentally demonstrated that amino acids could be created from electrical discharges, simulating lightning storms, within the gases of the primitive atmosphere of earth. The next important step came when Sidney Fox produced long-chain *proteinoids* (proteinlike molecules exceeding molecular weights of 10,000) by dry-heating of amino acids. Proteinoids have been shown to have characteristics of enzymes in that they catalyze some types of chemical reactions, and denature with mild heat and chemicals known to inhibit enzymes.

Two hypotheses have been proposed for the way in which cells formed. The first is through the condensation of *coacervate droplets*. Agitation of two or more polymers in water leads to the formation of these droplets. Coacervate droplets can be observed to accumulate and concentrate materials from the surrounding environment. Accumulated polymers are found to become stabilized within these droplets. Another way in which cells may have originally formed is through the formation of *proteinoids*. Fox found that when water was added to proteinoids, they formed dense *microspheres* that sank in water. It can be demonstrated that microspheres have osmotic properties and that the double layer of protein that serves as a boundary for each microsphere, is selective and readily admits polynucleotides (nucleic acid precursors).

The demonstrated activities of coacervates and microspheres may resemble rudimentary forms of metabolism but are a far cry from the numerous and complex reactions associated with a living cell. The giant step from nonliving droplet to living "droplet" (the cell) was probably the result of chance and trial and error over long periods of time. *Cells* are units of biological activity, surrounded by a semipermeable membrane, and capable of self-reproduction in a medium free of other living systems. In the development of the cell, DNA eventually became the repository for the genetic code and blueprint for proteins, while RNA became the means by which protein synthesis could be directed.

The first cells were probably *heterotrophs,* cells or organisms requiring preformed organic molecules. It is believed that the *autotrophs,* cells or organisms that can produce their own organic molecules from carbon dioxide, water, and sunlight, evolved later in time. This means that anaerobic respiration was the first form of respiration to appear. With the evolution of the autotrophs came a change in the atmosphere from one that was reducing to one that is currently oxidizing. As we have seen, a byproduct of photosynthesis is molecular oxygen.

The next large step was undoubtedly then the development of aerobic respiration. Aerobic respiration of organic molecules has a greater yield of energy for biological work than does either anaerobic respiration or photosynthesis. Somewhere along this line of evolution of the cell came the rise of the *eukaryotic cell,* which contained numerous membrane-bound organelles. Along with the development of the eukaryotic cell came the all-important ability to reproduce sexually. Sexual reproduction allows for variation through genetic shuffling.

THEORIES OF EVOLUTION

The two opposing, nonreligious theories of *evolution,* the rise by descent and modification of ancient forms into present-day organisms, are *Lamarckism* and *Darwinism.* Lamarck held that species evolved as a result of the characteristics acquired by one generation being inherited by the next generation. He used the giraffe as an example. He proposed that giraffes developed long necks as a result of generation upon generation of giraffes stretching their necks in attempts to reach higher and higher into the trees for food. Darwin, on the other hand, proposed that species evolved as a result of inheritance of characteristics of those individuals that best survived in nature. Current evidence supports Darwin's proposal. Had Darwin known how genes were inherited, he would have been able to refute Lamarck's proposal.

With the establishment of life on earth, the "seeds" for organismic evolution had been sown. Centuries before the concept of evolution was proposed, there was abundant evidence for evolution in the practices of animal and plant breeders. These individuals have practiced and continue to practice *selective* or *artificial breeding,* the selection of mating pairs that best represent desired traits. The current theory of evolution was first formally proposed by Charles Darwin and Alfred R. Wallace, in separate papers presented at the same meeting of the Linnaean Society of London in June of 1858. Darwin and Wallace had arrived independently at the same conclusions about evolution.

In modern terms, we can define evolution as changes in the *gene pool,* all of the genes present in a population of organisms, from one generation to the next. *Natural selection* plays an important role in the process of evolution, as each species has different survival and reproduction of genotypes from generation to generation. Darwin's theory of evolution by natural selection is based upon three sets of observations.

1. There is variation within a population of organisms.
2. Many of the differences between and among members of a species are inherited.
3. Females produce many more offspring than live to mature sexually and reproduce.

It is clear from these three observations that those members of a population that inherit traits favoring survival in nature are more likely to develop to sexual maturity; they are able to pass their traits on to the next generation. Three general types of natural selection operate within natural populations. These are *stabilizing selection,* the process that eliminates any extreme individuals in a population, *disruptive selection,* the process that increases the extreme individuals at the expense of the mainstream population, and *directional selection,* the process that increases the proportion of certain individuals over others with the gradual replacement of one or more alleles with others from the gene pool.

The first example of natural selection was provided more than 100 years after Darwin and Wallace proposed their theory of evolution. While studying moths and butterflies in England, Bernard Kettlewell noticed that in collections from the mid-1800s, the dark variety of the "peppered moth" was rarely caught but in collections from the late 1800s to the early 1900s, the light variety of the moth was rarely caught. He attributed this change from the dominance of the light variety to the dominance of the dark variety to an effect of industrial pollution. Prior to the industrial revolution and the extensive burning of coal, the lichens covering many tree trunks were light in color and would best camouflage the light variety of moth. The darker moths would be contrasted against the light background and would therefore be selected against more readily by predators. The burning of coal caused the accumulation of soot on tree trunks, eventually making the background dark, camouflaging the dark variety of moth and making the lighter moths visible to predators. This phenomenon is commonly called *industrial melanism* and it is an example of *evolutionary adaptation*.

In biological terms, adaptation can mean a state of adjustment to the environment (*Environmental adaptation*), a state of adjustment within the lifetime of an individual's functioning (*physiological adaptation*), or a state of adjustment in a population over the course of generations (*evolutionary adaptation*). Drug resistance in bacteria, the change in the dominant variety of peppered moth, and many other examples support the concept of evolutionary adaptation.

Modern evolutionists are using a two-pronged approach to the study of evolution. *Macroevolution*, large-scale evolutionary change originally interpreted on the basis of the fossil record, is being reinterpreted in light of continental drift theory and comparative biochemical studies of macromolecules such as DNA, cytochrome c, etc. *Microevolution*, gene-frequency changes in local populations, is another approach being used to understand how life on earth evolved. The four major patterns of evolution are *divergent evolution*, *convergent evolution*, *parallel evolution*, and *coevolution*. Divergent evolution is the result of one population of a species becoming isolated from the rest of the species and then following a different course of evolution as a result of selective pressures that are different from those imposed on the original species. The example of the brown bear and the polar bear is one of divergent evolution. At one point in history, the "prepolar" bear stock is thought to have become isolated from the main group of brown bears (mainly herbivores) and to have developed into present-day polar bears (mainly carnivores). Convergent evolution is the result of populations of different species being exposed to similar selective pressures and exhibiting similar adaptations. The example of whales and sharks is one of convergent evolution. Although quite different in many respects, the whales and sharks found themselves subject to similar selective pressures and have evolved many similar external features; another example would be that of the seal and penguin. Parallel evolution describes the situation where two ancestral species resemble one another's adaptations and the descendants of each of those species evolve along similar lines, neither diverging nor converging from one another's evolutionary patterns. The classic example of parallel evolution is the evolution of marsupials in Australia and the evolution of placental mammals in other parts of the world. Coevolution is the name applied to the situation of two or more populations that interact so closely that each is a strong selective force on the other; among the best examples are those of flowers and the insects that pollinate them.

Population genetics is the synthesis of the principles enunciated in Mendelian genetics and Darwinian evolution. A *population* is a naturally interbreeding group of organisms. In natural populations, some of the alleles are always changing in frequency. Favorable alleles and combinations of favorable alleles are likely to increase in frequency in succeeding generations, while unfavorable ones are likely to decrease in frequency. Direct determination of gene frequencies in a gene pool is difficult at best. It is, however, possible to determine what percentage of a population expresses particular genotypes and from these data to calculate allelic gene frequencies. The theoretical basis for such determinations is an equation that expresses the *Hardy–Weinberg principle*. The Hardy–Weinberg principle predicts the genetic outcome of random matings in an idealized population of diploid organisms that reproduce sexually. The following is a summary of the relationships between allele frequencies and genotype frequencies.

Frequency of the dominant allele $= p$
Frequency of the recessive allele $= q$

Allele frequencies $= p + q = 1$ (square both sides yields)
Genotype frequencies $= p^2 + 2pq + q^2 = 1$ (equilibrium equation)

The equilibrium equation for genotypic frequencies states that in large sexually reproducing populations where random matings occur, there is no change with respect to the proportion of genotypes or the frequencies of alleles in the gene pool. The principal cause of change in the Hardy–Weinberg equilibrium is differential reproduction, the result of natural selection. Other influential factors are mutation (a change in the organic base sequence of DNA), gene flow (migrations of individuals with new genes into a population), genetic drift (sampling errors), and, in rare cases, nonrandom matings. Mutation is a raw material for evolution, but as mutation rates are very low it is unlikely that they significantly influence gene frequencies.

MINIQUIZ ON EVOLUTION

1. Although Charles Darwin is the acknowledged "father" of the theory of evolution, credit must also be given to
   (A) Charles Lyell
   (B) Thomas Malthus
   (C) Robert Brown
   (D) Gregor Mendel
   (E) Alfred Wallace

2. It is believed that the primitive atmosphere of earth lacked
   (A) water vapor
   (B) molecular oxygen
   (C) ammonia
   (D) methane
   (E) molecular hydrogen

3. The most likely sequence of events in the origin of life on earth is
   (A) reducing atmosphere—simple organic molecules—complex organic molecules—coacervate droplets or microspheres—heterotrophic cells—autotrophic cells—oxidizing atmosphere
   (B) reducing atmosphere—oxidizing atmosphere—coacervate droplets or microspheres—complex organic molecules—autotrophic cells
   (C) oxidizing atmosphere—complex organic molecules—coacervate droplets or microspheres—reducing atmosphere—autotrophic cells
   (D) autotrophic cells—complex organic molecules—coacervate droplets or microspheres—cells—reducing atmosphere
   (E) reducing atmosphere—coacervate droplets or microspheres—cells—oxidizing atmosphere

4. Birds and bats would be the best example of
   (A) coevolution
   (B) parallel evolution
   (C) divergent evolution
   (D) convergent evolution
   (E) none of these

5. Population shifts observed between the light and dark varieties of peppered moth in England is an example of
   (A) coevolution
   (B) parallel evolution
   (C) divergent evolution
   (D) convergent evolution
   (E) none of these

6. The unit of biological activity that is surrounded by a semipermeable membrane, and is capable of self-reproduction in a medium free of other living systems is a/an
   (A) coacervate droplet
   (B) cell
   (C) virus
   (D) microsphere
   (E) proteinoid

7. The first experimental evidence in support of the concept of organic origins for life on earth came from the work of
   (A) James Watson
   (B) Joseph B.S. Haldane
   (C) Alexander Oparin
   (D) Charles Darwin
   (E) Stanley Miller

8. Whales and sharks are an example of
   (A) convergent evolution
   (B) divergent evolution
   (C) parallel evolution
   (D) coevolution
   (E) microevolution

9. In the equation $p^2 + 2pq + q^2 = 1$, the frequency of the homozygous recessive gene is represented by the term
   (A) p^2
   (B) q^2
   (C) $2pq$
   (D) $p^2 + 2pq$
   (E) $2pq + q^2$

10. Assume that a population of humans has been determined to contain 75% individuals who are tasters and 25% who are nontasters. Determine which gene is more common in the population.

Answers and Explanations for the Miniquiz

1. **(E)** also concluded that the variety of life on earth was the result of evolution by natural selection and reported his conclusions at the same meeting as Darwin did. (A) was a geologist who influenced and encouraged Darwin to pursue his studies into evolution. (B) influenced Darwin's thinking by having used the term "struggle for existence" in one of his writings. (C) was the first person to describe the nucleus of the cell. (D) is the father of genetics.

2. **(B)** did not appear in the atmosphere until autotrophs appeared and carried out the all-important process of photosynthesis.

3. **(A)** is correct. Check back in the text material of this section.

4. **(D)** is the result of populations of different species being exposed to similar selective pressures and exhibiting similar adaptations. In this case, the wing. (A) can be exemplified by the relationship between certain insects and plants. (B) can be exemplified by the relationship between Australian marsupials and placental mammals in other parts of the world. (C) can be exemplified by the relationship between brown bears and polar bears.

5. **(E)** is correct because the example given is one of evolutionary adaptation.

6. **(B)** is correct. This is the definition of a cell.

7. **(E)** was able to demonstrate that in the presence of electrical discharge, inorganic gases can be converted into simple organic compounds. (A) played an important role in determining the structure of DNA. (B and C) proposed that life on earth arose from simple molecules on the early earth. (D) proposed the theory of evolution by natural selection.

8. **(A)** is correct. See answer 4.

9. **(B)** is correct. (A) is the frequency of the homozygous dominant gene. (C) is the frequency of the heterozygous condition. (D and E) are merely sums of terms.

10. Both genes are equally common. See the explanation below: $q^2 = 0.25$, therefore $q = 0.5$, which is the frequency of the nontaster gene. $1.0 - 0.5 = 0.5$ is the frequency of the taster gene.

ORGANISMS AND POPULATIONS

TAXONOMY

Words To Be Defined

BINOMIAL NOMENCLATURE
TAXA
KINGDOM
PHYLUM
CLASS
ORDER
FAMILY
GENUS
SPECIES
VIRUS
MONERAN
PROCARYOTIC CELL
EUCARYOTIC CELL
PROTIST
FUNGUS
MONOTREME
MARSUPIAL
PLACENTAL MAMMAL
PLACENTA
ALTERNATIONS OF GENERATIONS
GAMETOPHYTE
SPOROPHYTE
SPORE
TRACHEOPHYTE
GYMNOSPERM
ANGIOSPERM

GENERAL OVERVIEW

The interactions of genetics and evolution over vast periods of time have led to an immense variety of living organisms adapted to almost any conceivable environment on the earth. Most estimates agree that there are some 2 million species on earth today; some estimates run as high as 10 million. Common names for life-forms are useful to residents of a specific region in which they are found, but are useless to scientists who try to determine relationships among life-forms. A system of *binomial nomenclature* (a system of classification using two names) was developed by Carolus Linnaeus in 1735. By convention, the first name is capitalized and either italicized or underlined (genus name). The second name is all in lower-case letters and either italicized or underlined (species name); e.g., the scientific name for man is *Homo sapiens. Homo* is the genus and *sapiens* is the species. Binomial nomenclature helps to avoid confusion because each organism receives a unique name.

CURRENT LEVELS OF CLASSIFICATION (Taxa)

Kingdom Any of the five major categories into which organisms are classifiied
Phylum The primary subdivision of a kingdom
Class The primary subdivision of a phylum of the plant or animal kingdom
Order The primary subdivision of a class of the plant or animal kingdom
Family The primary subdivision of an order of the plant or animal kingdom
Genus The primary subdivision of a family of the plant or animal kingdom
Species The primary subdivision of a genus of the plant or animal kingdom

Today, the classification of organisms employs five kingdoms: Monera, Protista, Fungi, Plantae, and Animalia. Although considered to have the attributes of life, *viruses* (submicroscopic, noncellular particles that consist of either DNA or RNA and a protein coat) are not members of any kingdom.

The kingdom **Monera** consists of members of the phylum Schizophyta (bacteria) and the phylum Cyanophyta (blue-green algae). All of these forms are composed of *procaryotic cells* (cells with no nuclei; chromosomes being suspended within the cytoplasm). All of the other kingdoms consist of forms composed of *eucaryotic cells* (cells with true nuclei enclosed within a nuclear membrane).

The kingdom **Protista** consists of many phyla. The key groups are

Euglenophyta—euglenoids
Chrysophyta—golden algae & diatomes
Zoomastigina—animal flagellates
Ciliophora—ciliates and suctorians
Sarcodina—rhizopods (Ameba)
Pyrrophyta—dinoflagellates
Sporozoa—sporozoans

The kingdom **Fungi** consists mostly of multicellular plantlike forms that all lack chlorophyll and therefore are heterotrophic. All fungi reproduce asexually by means of spores. The two phyla in the kingdom are:

Myxomycophyta—slime molds
Eumycophyta—true fungi, including the molds, yeasts, mushrooms, and ringworm fungi.

The kingdom **Plantae** is a large group of multicellular plants that includes the phyla

Rhodophyta—red algae
Chlorophyta—green algae
Lycopodophyta—club mosses
Pterophyta—ferns
Coniferophyta—conifers
Phaeophyta—brown algae
Bryophyta—liverworts, hornworts, and mosses
Arthrophyta—horsetails
Cycadophyta—cycads
Anthophyta—flowering plants

The kingdom **Animalia** is a large group of multicellular animals that includes the phyla

Mesozoa—mesozoans
Cnidaria—coelenterates
Platyhelminthes—flat worms
Acanthocephala—spiny-headed worms
Mollusca—mollusks
Arthropoda—arthropods
Hemichordata—acorn worms
Porifera—sponges
Ctenophora—comb jellies
Nemertinea—ribbon worms
Aschelminthes—pseudocoelomate worms
Annelida—segmented worms
Echinodermata—echinoderms
Chordata—chordates

A SUMMARY OF MAJOR ANIMAL PHYLA

PHYLUM	CLASS	CHARACTERISTICS	EXAMPLES
CNIDARIA (Coelenterata)	—	Two tissue layers, ectoderm & endoderm; specialized stinging cells in ectoderm; radial symmetry	—
	Hydrozoa	Polyp stage dominant; many forms mobil	Hydra
	Anthozoa	Polyp stage only; no medusa	Sea anemones & corals
	Scyphozoa	Medusa stage dominant;polyp reduced	Jellyfish
PLATYHELMINTHES	—	Three tissue layers, ectoderm, mesoderm, & endoderm; beginnings of cephalization; bilateral symmetry	—
	Turbellaria	Free-living flatworm	Planaria
	Trematoda	Parasitic; suckers or adhesive structures degenerative organs	Flukes
	Cestoda	Parasitic; scolex that produces proglottids; degenerative organs	Tapeworms
ANNELIDA	—	Segmented worms; coelomate; setae & parapodia in some; respiration by skin or gills; mouth-to-anus gut	—
	Polychaeta	Head usually with parapodia and setae; tube dwelling and free crawling	Bristle worms
	Oligochaeta	Mostly terrestrial; setae but no parapodia; head reduced	Earthworms Leeches
	Hirudina	Parasitic forms with degenerative organs; setae absent; body flattened dorsoventrally	
MOLLUSCA	—	Segmentation reduced; body covered by mantle; head & muscular foot usually present	—
	Amphineura	Eight-plate shell usually present; one head; foot reduced	Chitons
	Bivalva	Flattened shell with two valves; mantle forms siphon; head reduced	Scallops, & mussels
	Gastropoda	Asymmetrical coiled shell; some with shells reduced or absent; radula present	Slugs; limpets & snails
	Cephalopoda	Prehensile tentacles surround the head; shell internal, external or absent; siphon mediated jet propulsion; large camera-type eye	Cuttlefish, squid, & octopus
ARTHROPODA	—	Chitinous, jointed exoskeleton; segmented body with a hemocoel	—
	Arachnida	One or two main body parts; four pairs of walking legs; two pairs of mouth parts	Ticks, mites, & spiders
	Crustacea	Two or three main body parts; three or more pairs of walking legs	Shrimp, crabs, & lobsters
	Insecta	Three-segment body (head, abdomen, & thorax); usually three pairs of walking legs; many with pairs of wings	Insects
	Diplopoda	Paired segments covered by a single skeletal plate; each segment possesses two pairs of walking legs	Millipeds
	Chilopoda	First segment appendages modified into poison claws; other segments bear a single pair of appendages	Centipeds

A SUMMARY OF MAJOR ANIMAL PHYLA

Phylum	Class	Characteristics	Examples
ECHINODERMATA	—	Radial symmetry; water vascular system with tube feet; calcareous skeleton	—
	Crinoidea	Ciliated feeding tubes; branched arms; free-moving or stalked	Feather stars & sea lilies
	Asteroidea	Arms merging into disk; tube feet on arms; free-moving	Sea stars
	Echinoidea	Skeleton of fused plates; no arms but do possess tube feet; free-moving	Sea urchins & sand dollars
	Ophiuroidea	Thin arms marked off from disk; mainly sensory tube feet; free-moving	Brittle stars & serpent stars
	Holothuroidea	Reduced skeleton; tube feet for locomotion, some modified as mouth tentacles; free-moving	Sea cucumbers
CHORDATA	—	Notochord at some stage in life; dorsal nerve cord; endoskeleton; ventral heart	—
Subphylum Vertebrata	—	Vertebral column; tail common; all have some sort of liver, kidney, and endocrine organ; pronounced cephalization	—
	Agnatha	Persistent notochord; cartilagenous skeleton; jawless	Lampreys & hagfish
	Chondrichthyes	Cartilagenous skeletons; paired jaw; paired pelvic and pectoral fins; tailfin usually asymmetrical	Sharks & rays
	Osteichthyes	Bony skeletons; gills covered by an operculum; paired pelvic and pectoral fins; tailfin usually symmetrical; many possess swimbladder; scales on body	Bony fishes
	Amphibia	Skin and lung respiratory mechanisms; egg laying without amnion or shell; no scales; tetrapods	Salamanders & frogs
	Reptilia	Lung respiration; egg laying with amnion and shell; scales on body; tetrapods	Snakes, turtles, alligators
	Aves	High body temperature; feathers; forelimbs modified into wings	Birds
	Mammalia	Mammary gland milk nourishment of young; body covered with hair; only one bone in each side of lower jaw; most viviparous	Mammals

There are three types of mammals. The *Monotremes* are egg-laying mammals, being represented by the duck-billed playtpus and the spiny anteater. The young hatch from shelled eggs and nurse from the mother's mammary glands. The *marsupials* are the pouched mammals. The young are born in an immature condition and must complete their development in the pouch on the body of the mother, where they nurse from the mother's mammary glands. The *placental mammals* are the most modern of the mammals. The eggs contains little yolk. All young initially develop in association with a *placenta*, a mass of blood vessels and membranes in the uterus which function in the exchange of materials between the embryo and the mother. The young are born at various stages of development. The mammals among all of the animals have developed the highest degree of care for the young.

HIGHER VASCULAR PLANTS

Early in the evolution of plants, there developed a pattern of *alternation of generations*. The alternation occurs between multicellular haploid individuals and multicellular diploid individuals. In the "lower" plants, the pattern is one of a dominant, independent, haploid (*gametophyte*) generation in which haploid individuals produce haploid gametes, without the benefit of meiosis, alternating with a usually reduced, dependent, diploid (*sporophyte*) generation in which diploid individuals produce haploid *spores* (reproductive haploid cells that do not undergo fertilization). In the "higher" plants the sporophyte generation becomes the independent, dominant generation, and the gametophyte generation becomes reduced to nothing more than a few cells, totally dependent on the sporophyte generation. The alternation of generations with similar or different body shapes, is characteristic of many plants. The *tracheophytes* (plants with organized vascular tissues) are the higher plants in evolution and include the ferns and the seed-producers. The *gymnosperms* produce "naked seeds" that are borne in ovules in reproductive structures called cones. The *angiosperms* produce seeds with protective coats. The seeds are borne in ovules that are enclosed in ovaries. The flower is the reproductive structure unique to the angiosperms.

MINIQUIZ ON TAXONOMY

Use the following key to select the correct answers for questions 1–5

(A) phylum Cnidaria (Coelenterata)
(B) phylum Annelida
(C) phylum Mollusca
(D) phylum Echinodermata
(E) phylum Chordata
(F) none of these

1. Includes spiders, crabs, millipeds, and insects

2. Includes snakes, bony fishes, lampreys, and cows

3. Includes hydra, corals, jellyfish, and sea anemones

4. Includes clams, mussels, squid, and octopus

5. Includes blood and liver flukes, tapeworms, and planaria

Use the following key to select the correct answers for questions 6–10

(A) class Anthozoa
(B) class Cephalopoda
(C) class Insecta
(D) class Osteichthyes
(E) class Amphibia
(F) none of these

6. Bony skeleton; gill covered by an operculum; paired pelvic and pectoral fins; tail usually symmetrical; scales over the body surface

7. Persistent notochord; cartilaginous skeleton; no lower jaw

8. Lung respiration; egg laying with amnion and a shell; scales over the body surface; tetrapod

9. Segmented into head, abdomen, and thorax; uses three pairs of walking legs; many forms with paired wings

10. Prehensile tentacles surrounding the head; shell internal, external, or absent; siphon-mediated jet propulsion; large camera-type eye

11. The system of binomial nomenclature used in taxonomy was first proposed by

    (A) Charles Darwin
    (B) Robert Brown
    (C) Stanley Miller
    (D) Carolus Linnaeus
    (E) Camillo Golgi

12. The taxonomic level between class and family is

    (A) kingdom
    (B) order
    (C) genus
    (D) species
    (E) phylum

Use the following key to select the correct answers for questions 13–17

(A) Euglenophyta, Ciliophora, and Eumycophyta
(B) Ciliophora, Pyrrophyta, and Phaeophyta
(C) Chrysophyta, Sarcodina, and Sporozoa
(D) Lycopodophyta, Coniferophyta, and Cycadophyta
(E) Euglenophyta, Ciliophora, and Ctenophora

13. Three phyla of the kingdom Protista are

14. Three phyla of the kingdom Plantae are

15. Two phyla of the kingdom Protista and one phylum of the kingdom Plantae are

16. Two phyla of the kingdom Protista and one phylum of the kingdom Animalia are

17. Two phyla of the kingdom Protista and one phylum of the kingdom Fungi are

18. The specific group of plants that produce flowers are known as

    (A) gymnosperms (Coniferophytes)
    (B) angiosperms (Anthophytes)
    (C) rhodophytes
    (D) pterophytes
    (E) bryophytes

19. The evolutionary pattern seen among the higher vascular plants is the change from

    (a) a dominant sporophyte generation to a dominant gametophyte generation
    (B) a dominant gametophyte generation to a dominant sporophyte generation
    (C) seeds for reproduction to spores for reproduction
    (D) a dependence on vascular tissues to a nondependence on vascular tissues
    (E) a dominance of flowers for reproduction to a dominance of cones for reproduction

20. The sporophyte generation in higher plants

    (A) produces haploid gametes
    (B) does not carry out meiosis
    (C) produces haploid spores that do not undergo fertilization
    (D) produces diploid spores that undergo fertilization
    (E) is the haploid generation

Answers and Explanations for the Miniquiz

1–5. **Phylum Cnidaria** (Coelenterata)—Polyps and jellyfish, animals with radial symmetry that are composed of two layers of cells. Bodies of a jelly-like consistency.
Phylum Annelida—Segmented worms with a well-developed coelom, one-way digestive tract, head, nephridia, closed circulation, and a well-defined nervous system.
Phylum Mollusca—Unsegmented coelomate animals with variously modified muscular foot, head, and mantle.
Phylum Echinodermata—Starfish and sea urchins, animals with radial symmetry in the adult stage. Well-developed coelom, exoskeleton of spines and ossicles, and a unique water vascular system.
Phylum Chordata—Animals with a notochord, pharyngeal gill slits, a hollow dorsal nerve cord, and a tail at some stage of their lives.

1. (**F**) 2. (**E**) 3. (**A**) 4. (**C**) 5. (**F**)

6–10. **Class Anthozoa**—Sea anemones, colonial corals, and related forms. Commonly called flower animals.
Class Cephalopoda—"Head-foot" animals with numerous arms, or tentacles, around a mouth with two horny jaws, a well-developed nervous system, and a well-developed camera-like eye.
Class Insecta—Body of three distinct parts (head, thorax, and abdomen). Compound eyes, one pair of antennae, three pairs of legs.
Class Osteichthyes—Bony fishes.
Class Amphibia—Larval forms breathe by means of gills, while adults use lungs. Naked skin and incomplete double circulation.

6. (**D**) 7. (**F**) 8. (**F**) 9. (**C**) 10. (**B**)

11. (**D**) is correct. (A) proposed the theory of evolution by natural selection. (B) is credited with applying the term "nucleus" to the center of the cell. (C) performed the first experiments that demonstrated that it is possible for life to evolve from conditions of the primitive earth. (E) is credited with being the first to identify the cellular organelle that serves as the cell's packaging center.

12. (**B**)

13. (**C**)

14. (**D**)

15. (**B**)

16. (**E**)

17. (**A**)

18. (**B**)

19. (**B**) is correct. (A, C, D, and E) are all correct if you reverse each.

20. (**C**) is correct. (A) Sporophytes produce haploid spores, not gametes. (B) Sporophytes do carry out meiosis in order to produce the haploid spores. (D) Spores are not diploid. (E) The sporophyte is the diploid generation.

ANATOMY AND PHYSIOLOGY OF PLANTS

VASCULAR PLANT ORGANS

Words To Be Defined

ROOT
SHOOT
STEM
LEAF
INDETERMINATE GROWTH
MERISTEM
APICAL MERISTEM
PRIMARY GROWTH
LATERAL MERISTEM
SECONDARY GROWTH
ANNUAL
BIENNIAL
PERENNIAL
HERBACEOUS
WOODY
MONOCOT
DICOT
COTYLEDON
EPIDERMIS
ROOT HAIR CELL
CORTEX
PARENCHYMA CELL
ENDODERMIS
STELE
PERICYCLE
XYLEM
PHLOEM
CUTICLE
PITH
PARENCHYMA
VASCULAR BUNDLE
COLLENCHYMA
VASCULAR CAMBIUM
WOOD
CORK CAMBIUM
CORK (OUTER BARK)
LENTICEL
SUBERIN
TWIG
TERMINAL BUD
LATERAL BUD
LEAF SCAR
BUD SCALE SCAR
STOMATA
GUARD CELL
MESOPHYLL
SPONGY MESOPHYLL
PALISADE MESOPHYLL
BLADE
TRACHEID
VESSEL
FIBER
SCLEREID
CAPILLARITY
ROOT PRESSURE
TRANSPIRATION
SIEVE CELL
SIEVE PLATE
COMPANION CELL
PHLOEM FIBER
MASS FLOW (PRESSURE FLOW, SOLUTION FLOW)

The vascular plants include the ferns, gymnosperms, and angiosperms; they exhibit, both a root system and a shoot system. The *root* is an organ of anchorage, food storage, and absorption (water and minerals). The *shoot* is made up of one or more stems with leaves. *Stems* are organs that transport substances between root and leaves, produce food if there are chlorophyll-containing cells present, store foods, and hold the leaves in such a position as to optimize exposure to sunlight. *Leaves* are the organs in which photosynthesis occurs.

The higher plant shows *indeterminate growth,* the proliferation of unspecialized cells that have the potential to form new organs throughout the plant's life. Clusters of the unspecialized cells are called *meristems. Apical meristems* are responsible for *primary growth,* principally growth in length of roots and stems; *lateral meristems* are responsible for *secondary growth,* principally growth in girth, or width of roots and stems.

The higher plants exhibit three growth patterns: *Annual* plants grow for one year, reproduce, and then die; *Biennial* plants grow for two years, reproduce the second year, and then die; *Perennial* plants grow for many years, reproduce for many years, and then die. One of the oldest perennials in the United States is over 5,000 years old. Annuals and biennials are usually *herbaceous,* vascular plants with small, soft, nonwoody bodies. Perennials are usually *woody,* vascular plants with large, hard, woody bodies.

Among the angiosperms there are those that are *monocots* and *dicots.* The monocots have seeds with only one *cotyledon,* a seed leaf that serves as a food reserve for the embryo, while the dicots have seeds with two cotyledons. Other differences between monocots and dicots are listed in the chart below.

COMPARISON OF MONOCOTS AND DICOTS

CHARACTERISTIC	MONOCOTS	DICOTS
LEAF VEIN PATTERN	Parallel vein pattern	Net vein pattern
GROSS LEAF STRUCTURE	Blade only	Blade with petiole
MICROSCOPIC LEAF STRUCTURE	Spongy mesophyll only	Spongy and palisade mesophyll
GROWTH PATTERN	Usually primary only (only primary meristems)	Both primary and secondary (both primary and secondary meristems)
TYPE OF ROOT SYSTEM	Fibrous root (no main root)	Taproot (one or two main roots)
FLOWER PARTS	In 3s, multiples of 3s, or irregular	In 4s, 5s, multiples of each, or irregular
FORMS	Mostly herbs; few trees	Herbs, shrubs, trees

Roots

The primary structure of the dicot root is seen in both cross and longitudinal section in Fig. 3–12. The outermost layer, the *epidermis* includes specialized cells called *root hair cells.* The root hairs increase the surface area available for absorption of water and nutrients in the soil. The *cortex,* several layers of cell beneath the epidermis, is composed of *parenchyma cells.* The cortex cells serve as storage areas for carbohydrates. The inner edge of the cortex is bounded by a single layer of cells that constitutes the *endodermis.* The endodermis controls the movement of substances between the root cortex and the central vascular-containing region

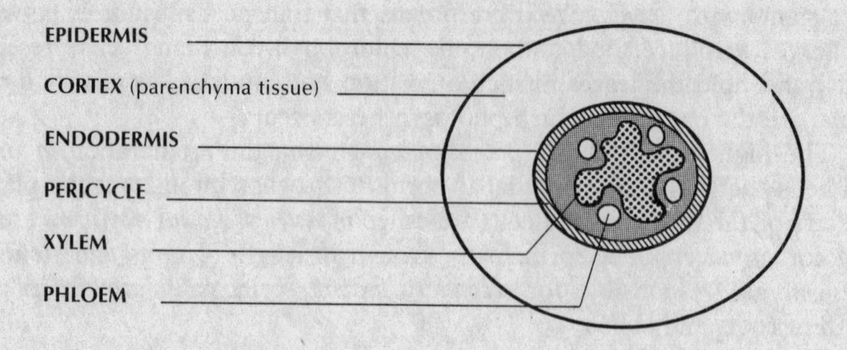

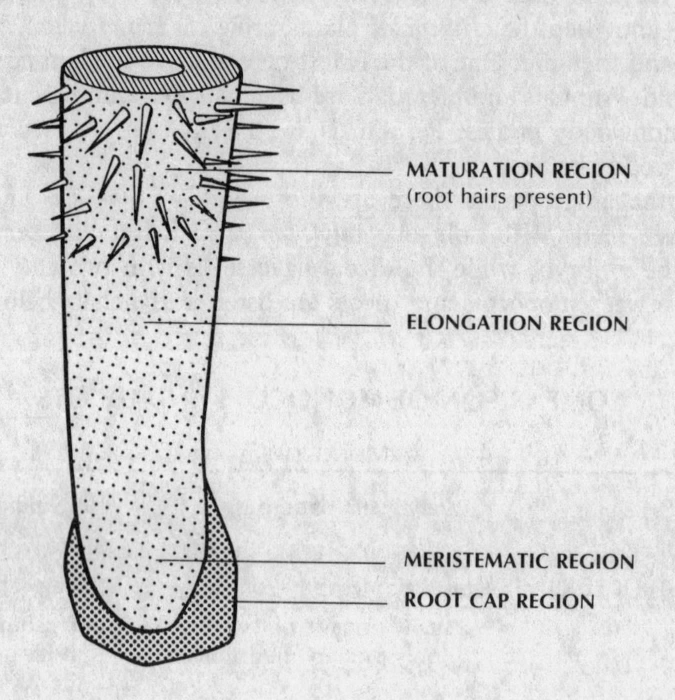

Fig. 3-12 Cross Section and Longitudinal Portion of a Typical Dicot Root Tip

called the *stele*. Just within the endodermis is the single-layered *pericycle*. The pericycle remains meristematic and produces new cells when needed. The center of the stele is filled with *xylem*, the tissue that transports water and minerals from the root, through the stem, and into the leaves. Alternating between the "arms" of xylem are patches of the tissue known as *phloem*. Phloem conducts food manufactured in the leaves to other parts of the plant.

Stems

Monocot stem growth is the result of activity of the apical meristem. New cells on the surface of the developing stem often produce a waterproof, waxy *cuticle* that is continuous with the cuticle that covers leaves. Beneath the epidermis is an extensive *pith* region composed of *parenchyma* cells around *vascular bundles*. These bundles are made up of *collenchyma*, column-shaped cells with thickened walls that help in the support of soft stems, xylem, and phloem, Fig. 3–13.

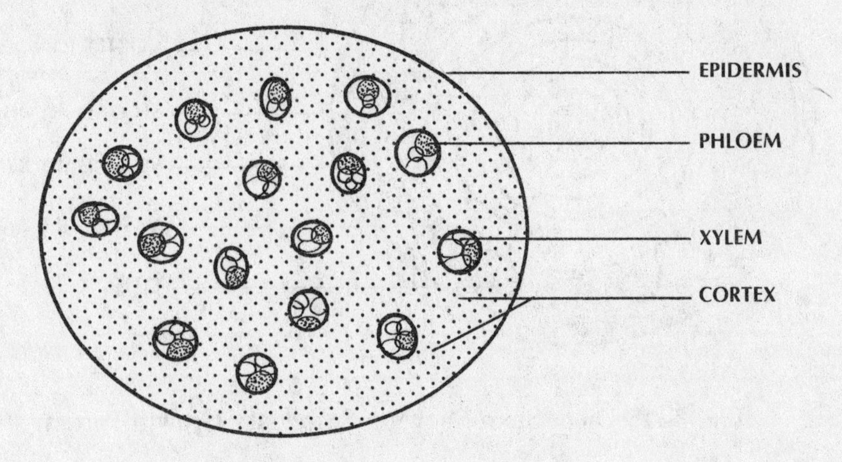

Fig. 3-13 Monocot Stem Showing Primary Growth Pattern

Primary dicot stem growth is the result of activity of the apical meristem. New cells on the surface of the developing stem often produce a waterproof, waxy cuticle similar to that found on the monocot stem and it too is continuous with the cuticle that covers the leaves. Beneath the epidermis is a cortex composed of collenchyma. The remainder of the cortex is composed of parenchyma. The *pith* in the center of the stem is also composed of parenchyma. The primary xylem and phloem form a ring of bundles toward the outer edge of the stem (see Fig. 3–14). As time goes on, the *vascular cambium*, a lateral meristem, forms and begins developing "layers" of xylem internally and phloem externally (see Fig. 3–15). Each year, a new layer of xylem is added and remains in the stem, while a new layer of phloem is added but begins to crush older layers of phloem. Secondary xylem forms the *wood* in the stem.

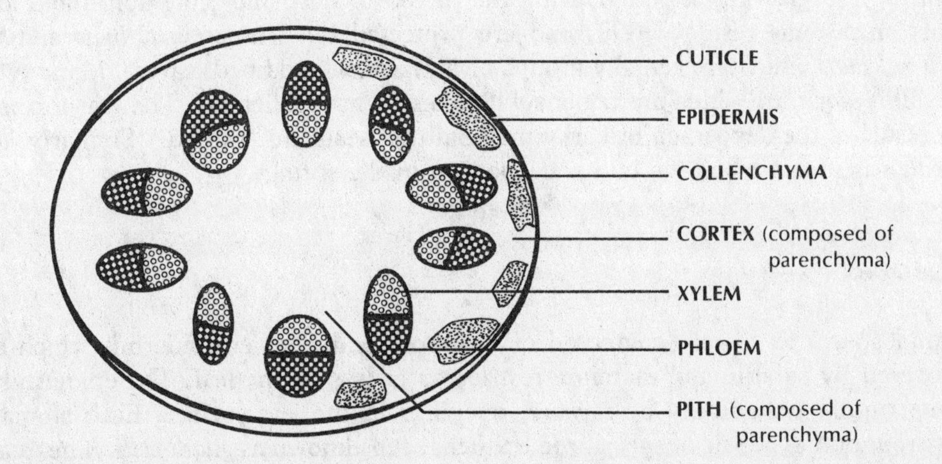

Fig. 3-14 Dicot Stem Showing Primary Growth Pattern

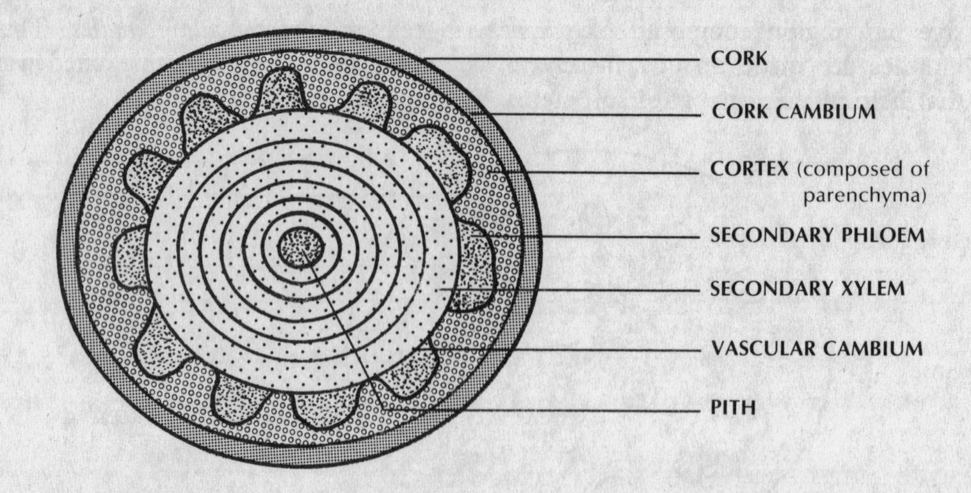

Fig. 3–15 Dicot Stem Showing Secondary Growth Pattern

Beneath the epithelium in older stems, a meristematic tissue known as the *cork cambium* forms and is responsible for the formation of *cork* or *outer bark*. The outer bark often exhibits what appears to be longitudinal cracks or *lenticels;* these are regions of gas exchange between the cells of the cortex and the external environment. The outer region of the bark often becomes impregnated with a waxy material known as *suberin*. The suberin prevents water loss through the stem. Secondary xylem layers can be used to determine such things as the age of a stem, the relative amount of rainfall from year to year, and competition between one tree and another.

Twigs are the finer branches of a tree. They bear leaves, flowers, and fruits during the growing season. During the period of dormancy, the apical and axillary meristems reside within, and are protected by, the *terminal buds* and the *lateral buds*. Buds are actually groups of bud scales (modified leaves). Many twigs exhibit *leaf scars* which are regions where leaves have fallen off. The scar forms as a result of the deposition of corky material that seals the "wound." Similarly, *bud scale scars* form when the bud scales fall off in the spring.

Leaves

Angiosperm leaves are composed of an upper and a lower epidermis which are covered by suberin and therefore restrict water loss by the leaf. The epidermis is interrupted periodically by *stomata*, openings in the leaf surface. Each stoma is surrounded by a pair of specialized epithelial cells known as *guard cells*. The guard cells regulate the size of the openings in the surface of the leaves. *Mesophyll*, photosynthetic tissue, is found between the epidermal layers. Scattered through the mesophyll are the vascular bundles which are composed of xylem, phloem, and collenchyma (see Figs. 3–16 and 3–17).

Monocot leaves differ from dicot leaves in several ways. First, monocot leaves usually do not have petioles while dicot leaves do. Second, the pattern of vascular bundles in the monocot leaf is parallel while the pattern in the dicot leaf forms a net. Finally, monocot leaves exhibit a *spongy mesophyll,* several layers of loosely arranged photosynthetic cells, while dicot leaves exhibit both a spongy mesophyll and a *palisade mesophyll,* one or two layers of long, thin, closely packed photosynthetic cells. The broad, flat part of a leaf is called the *blade.*

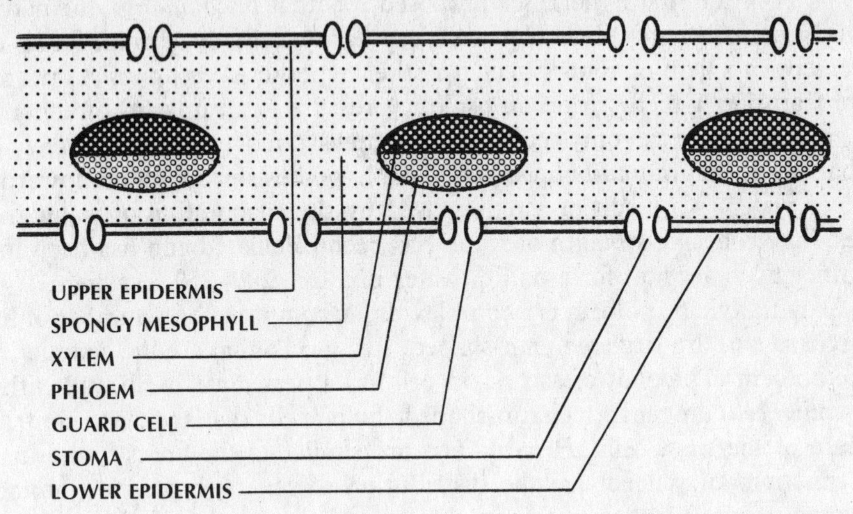

Fig. 3-16 Monocot Leaf Structure

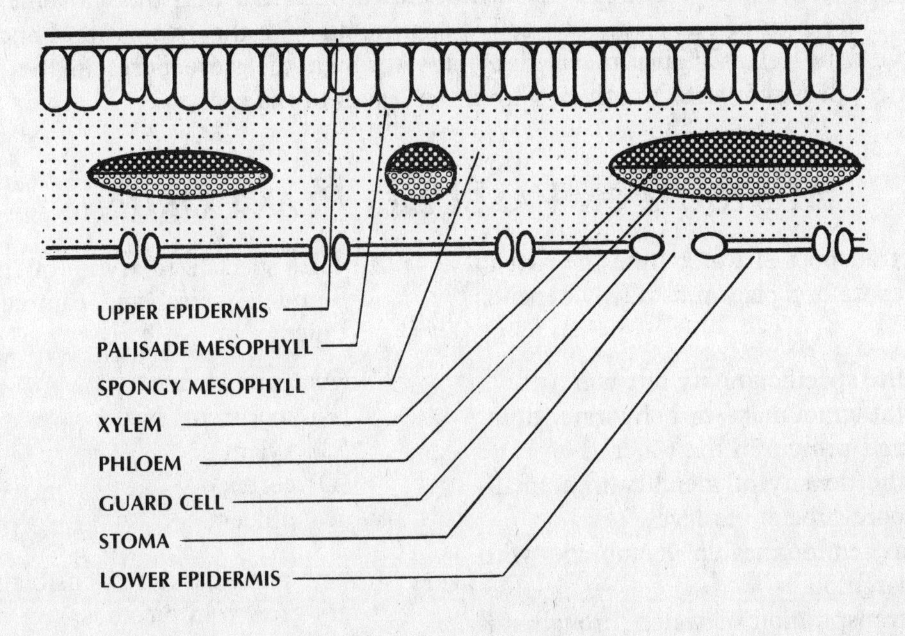

Fig. 3-17 Dicot Leaf Structure

Vascular Plant Transport

Xylem is a tissue made up of a number of kinds of cells. *Tracheids* are long, thin, thick-walled, dead cells that transport water and minerals. *Vessels* are short, wide, thick-walled, dead cells that also transport water and minerals. *Fibers,* long, thick-walled, often dead cells, and *sclereids,* variable-shaped, thick-walled, dead cells, are both used to strengthen xylem. Parenchyma is composed of relatively thin-walled, relatively unspecialized living cells that store starch and carry out lateral conduction. The conduction of fluids in xylem can be explained in several ways. One of the first hypotheses proposed was that of *capillarity,* the natural rise of fluids in small-bore tubes. However, capillarity can only account for an upward movement in xylem of about 5 feet. Another hypothesis, *root pressure,* proposes an active transport which creates an osmotic pressure in an upward direction. However, root pressure can only account for an upward movement of 1–2 feet. A third mechanism is that of *transpiration pull.* This hypothesis proposes an upward movement of water as a result of a pull caused by the evaporation of water from the leaves. The cohesive strength of water has been calculated and has been found to be sufficient to account for a pull of water in a tree 200–250 feet high.

As with xylem, phloem is a complex tissue composed of a number of kinds of cells. *Sieve cells* are large, column-shaped, living cells that are the conducting cells of the phloem. The ends of each sieve cell has a *sieve plate,* a cell wall with pores. *Companion cells* are small, column-shaped, living cells that appear to contribute to the care of the sieve cells. *Phloem fibers* are similar to the fibers found in xylem. The functions of phloem are the distribution of sugars, hormones, amino acids and some minerals to flowers, fruits, buds, stems, and roots. Phloem transport is still poorly understood. The most widely accepted theory of phloem transport is that of *mass flow* (also called *pressure flow,* or *solution flow*). This theory contends that osmotic pressures force the mass of solutes in water to move along a phloem gradient for mass. There are no membranes to interfere with this movement as the cytoplasm of one sieve tube cell is continuous with the cytoplasm of another sieve tube cell. Cytoplasmic bridges pass through the sieve pores in the sieve plates, providing direct pathways from the leaves to the roots.

MINIQUIZ ON VASCULAR PLANT ORGANS

1. The transport of water from the root to the leaves of a plant most likely depends on

   (A) the specific gravity of water
   (B) the amount of carbohydrate, lipid, and protein in the water
   (C) the density of mercury in a small-bore tube at sea level
   (D) a semipermeable membrane with large pores
   (E) transpiration of water through stomata in leaves

2. The specialized type of plant tissue found in apical and cambial regions is called

   (A) parenchyma
   (B) meristem
   (C) xylem
   (D) cortex
   (E) phloem

3. One plant tissue responsible for secondary growth in dicots is

   (A) vascular cambium

(B) pericycle
(C) apical meristem
(D) endodermis
(E) lateral bud

4. The complex plant tissue composed of parenchyma, sclereids, specialized conducting cells, and companion cells is

(A) cambium
(B) phloem
(C) cortex
(D) endodermis
(E) xylem

5 Leaf epidermal cells often secrete a waxy waterproofing material known as

(A) suberin
(B) cerumen
(C) exudate
(D) chlorophyll
(E) sap

6. One characteristic of monocot plants is

(A) a taproot system
(B) net-veined leaves
(C) a leaf with a petiole
(D) a palisade mesophyll in the leaf
(E) no secondary growth

7. The gas exchange opening in the leaf is the stoma, while the gas exchange opening on the stem is the

(A) pericycle
(B) root hair
(C) lenticel
(D) cork cambium
(E) xylem

8. The photosynthetic tissue in plants is called

(A) collenchyma
(B) xylem
(C) epidermis
(D) endodermis
(E) mesophyll

9. Scattered vascular bundles surrounded by parenchyma is characteristic of a

(A) monocot leaf
(B) dicot root
(C) dicot stem
(D) monocot stem
(E) dicot leaf

10. The best explanation of water transport from the roots to the leaves in tall trees is

(A) an upward push of water caused by the development of root pressure in response to osmosis
(B) the upward mass flow of solutes in cytoplasm of the xylem
(C) the natural capillarity in the small-bore xylem tubes
(D) a pull of water upward in response to the transpiration of water from the leaves of tall trees
(E) the diffusion of water across semi-permeable membranes.

11. The region of a plant that serves as a storage area for carbohydrates is the

(A) pericycle
(B) stele
(C) cortex
(D) epidermis
(E) cork

Answers and Explanations for the Miniquiz

1. **(E)** is correct. (A) relates to what is dissolved in water. (B) relates to the solutes in water. (C) is nonsense. (D) would not relate because functional xylem is dead.

2. **(B)** is correct. (A) is a tissue used for storage of carbohydrate. (C) is the tissue for water and mineral conduction from the roots to the leaves. (D) is the region of parenchyma. (E) is the tissue for food transport.

3. **(A)** is one of the lateral cambia. (B) is capable of becoming meristematic and forming the apical meristem of a lateral root. (C) is

responsible for primary growth. (D) is not responsible for any kind of growth. (E) is composed of leaf scales that cover an apical meristem.

4. **(E)** is as described in the question. (A) is a meristematic tissue. (B) is composed of sieve cells, companion cells, and fibers. (C) is composed primarily of parenchyma. (D) is a unique tissue.

5. **(A)** is correct. (B) is ear wax in mammals. (C) is a general cellular secretion. (D) is the photosynthetic pigment in green plants. (E) is the fluid in xylem.

6. **(E)** is correct. (A, B, C, and D) are all true of dicot plants.

7. **(C)** is correct. (A) is a region that may focally develop into an apical meristem in lateral root development. (B) provides increased surface area for absorption of water and minerals from the soil. (D) is a lateral meristem for the development of cork. (E) is the tissue that conducts water and minerals from the roots to the leaves.

8. **(E)** is the chlorophyll-containing tissue of the leaf. (A) is a support tissue. (B) is a conducting tissue. (C) is a nonphotosynthetic tissue covering leaves and young stems. (D) is a unique tissue.

9. **(D)** is correct. (A) does not contain parenchyma. (B) does not contain scattered vascular bundles. (C) has organized vascular bundles. (E) has scattered vascular bundles in mesophyll, not parenchyma.

10. **(D)** is correct. (A and C) can only account for upward movements of 1–2 feet. (B and E) are nonsense.

11. **(C)** is correct. (A) is the region of lateral root development. (B) is the core of vascular tissue in the root. (D) is the tissue that covers leaves and young stems. (E) is the dead protective layer which replaces epithelium in older dicot stems.

PLANT HORMONES

Words To Be Defined

TROPISM
GEOTROPISM
PHOTOTROPISM
HYDROTROPISM
POSITIVE TROPISM
NEGATIVE TROPISM
SHORT-DAY PLANTS
LONG-DAY PLANTS
DAY-NEUTRAL PLANTS

Plant hormones are often produced in tissues that are responsible for other functions. In animals, hormones are produced in tissues whose main function is the production of the hormone. Unlike animal hormones, plant hormones seem to affect virtually any plant tissue. Hormones in plants govern growth in such a way as to balance the anatomy and physiology of the plant. Plants do not have nervous systems, and hormones, therefore, also provide the mechanism by which plants respond to environmental stimuli. The general effect is an irreversible, permanent change in a tissue; e.g., cell division, cell elongation, and cell death. Hormones enable a plant to respond to the direction of such environmental factors as gravity, light, or prevailing winds. Hormones can also be used to measure the length of daylight through seasonal changes. Various *tropisms,* growth responses to environmental factors, are under the control of hormones. Some examples of tropism are *geotropism,* response to gravity, *phototropism,* response to light, and *hydrotro-*

pism, response to water. If the plant moves toward the environmental factor, it is identified as *positive tropism.* If a plant moves away from an environmental factor, it is identified as *negative tropism.* Plant response to a hormone depends on tissue type, hormone concentration, interaction of two or more hormones present in the same tissue, tissue physiological condition, tissue age, and environmental factors. The chart below summarizes the six groups of hormones that have been identified and studied.

HORMONE GROUP	FUNCTIONS	MECHANISM
AUXINS	Positive phototropism in stems; induction of cell division and lateral bud growth (*apical dominance*); stimulation of adventitious root, lateral root, female flower part, and fruit development	Stimulation of cell elongation in stems; inhibition of cell elongation in the cambium
GIBBERELLINS	Stimulation of growth of stems and leaves; stimulation of cell division of stem apex; stimulation male flower part development	Stimulation of cell elongation
CYTOKININS	Stimulation of plant growth and the production of fruits and seeds; prevention of the onset of dormancy; slowing the process of aging	Stimulation of cell division
ABSCISIC ACID	Stimulation of aging and mediation of positive geotropic response by roots; promotion of flowering, leaf dropping, and formation of buds in short-day plants	Causes the shutdown of the synthesis of RNA and protein
ETHYLENE	Inhibition of the effects of auxins; stimulation of aging in cells; and stimulation of ripening in fruits	Stimulation of the production of cellulase, an enzyme that breaks down cellulose; stimulation of the production of other enzymes involved in ripening processes.
PHYTOCHROME	Stimulation of flowering in response to length of the dark period over 24 hours.	Stimulation of local changes in the electrical potential of cell membranes.

Flowering in many plants appears to require a critical period of darkness each day. Plants that require a long period of darkness each day in order to flower are called *short-day plants.* Plants that require a short period of darkness each day in order to flower are called *long-day plants. Day-neutral plants* seem to flower in response to age and are not influenced by the length of darkness each day. Short-day treatment stimulates the production of abscisic acid which promotes flowering in short-day plants. Experimental evidence shows that the application of abscisic acid to short-day plants held under long-day conditions will cause them to flower. This seems to suggest that the correct concentration of abscisic acid can substitute for short-day treatment. Similarly, experimental evidence shows that the application of gibberellic acid to long-day plants held under short-day conditions will cause them to flower. The correct concentration of gibberellic acid can substitute for long-day treatment.

MINIQUIZ ON PLANT HORMONES

Use the key below to answer questions 1–5

(A) auxins
(B) gibberellic acid
(C) cytokinins
(D) abscisic acid
(E) ethylene
(F) phytochrome

1. The hormone that can substitute for short nights in stimulating flowering in long-day plants is

2. The hormone that counteracts the effects of auxins is

3. The hormone that can substitute for long nights in stimulating flowering in short-day plants is

4. The hormone that stimulates the ripening of fruit is

5. The hormone that is the hormone of flowering is

6. One generalization that can be made about plant hormones is that they

   (A) stimulate cell elongation
   (B) maintain homeostasis in plants
   (C) promote flowering
   (D) inhibit RNA and protein synthesis
   (E) inhibit cell division

7. The movement of a plant root toward the earth is known as

   (A) negative geotropism
   (B) positive phototropism
   (C) negative hydrotropism
   (D) positive geotropism
   (E) none of these

Answers and Explanations for the Miniquiz

1. **(B)**

2. **(E)**

3. **(D)**

4. **(E)**

5. **(F)**

6. **(B)** is correct. (A, C, D, E) are not generalizations.

7. **(D)** is correct. (A) is movement away from the earth. (B) is movement toward light. (C) is movement away from water.

PLANT REPRODUCTION

Words To Be Defined

VEGETATIVE REPRODUCTION
SEXUAL REPRODUCTION
ZYGOTE
SEPAL
PETAL
STAMEN
CARPEL
FILAMENT
ANTHER
STIGMA
STYLE
OVARY
POLLEN GRAIN
SPORE
MEGASPORE
OVULE
ENDOSPERM NUCLEUS
DOUBLE FERTILIZATION
FRUIT
RECEPTICLE
SUSPENSOR
ENDOSPERM
RADICLE
PLUMULE
COTYLEDON

The two main types of reproduction in plants are *vegetative reproduction,* extended growth giving rise to new individuals that are genetically identical to the parent plant, and *sexual reproduction,* the fusion of haploid male cells with haploid female cells and followed by growth giving rise to new individuals that resemble, but are genetically different from either of the two contributing parents. Sexual reproduction in angiosperms requires a complex sequence of events. First, meristematic cells must differentiate to become flower parts. Select cells within the flower must undergo meiosis to produce male and female gametophytes. Then female gametophytes must form egg nuclei while male gametophytes must form sperm nuclei. The egg and sperm cells must join in the process of fertilization in the formation of the *zygote,* the first diploid cell of the new sporophyte generation. The zygote must then develop into an embryo with stored food and a seed coat.

The reproductive structure in angiosperms is the flower, which is actually a modified shoot. A typical flower is composed of: *sepals,* a ring of green, outer, basal leaves; *petals,* a ring of modified leaves within the sepals, often bearing colored patterns that attract animal pollinators; *stamens,* modified leaf structures that bear the male reproductive organs; and an innermost ring of *carpels,* modified leaf structures that bear the female reproductive organs. The stamen is further subdivided into a long stalk, the *filament,* and a chamber where pollen develops, the *anther*. Similarly, the carpel is divided into a structure that receives the pollen, the *stigma;* a stalk that connects the stigma to the ovary, the *style;* and the structure that encloses one or more ovules, the *ovary*.

The *pollen grain* is the male reproductive cell that develops within the anther of the flower. Microspore mother cells in an anther undergo meiosis, producing four haploid *spores* for each cell. The nucleus of each spore divides by mitosis, producing a haploid cell with two nuclei. This cell is then called a pollen grain. The *megaspore* is the female reproductive cell that develops within the *ovule* of the ovary. Megaspore mother cells in an ovule undergo meiosis, producing four haploid megaspores for each cell. Fertilization occurs anywhere from one hour to several months after pollination. The angiosperms are unique in that each pollen grain possesses two sperm nuclei. One haploid sperm nucleus fuses with the haploid egg nucleus to form a diploid zygote and the other haploid sperm nucleus fuses with two central haploid nuclei to form a triploid nucleus known as the *endosperm nucleus*. For this reason, flowering plants are said to have a *double fertilization*.

Once the ovule is fertilized, the zygote begins developing into an embryonic plant with the parent plant supplying nutrients until the embryo becomes established as an individual. During this process, the ovule develops into a seed coat and the ovary turns into a *fruit*. In some plants, the tip of the flower stalk, the *receptacle*, turns into the fruit. After fertilization there is a short period of dormancy followed by the first stage of development. The zygote divides by mitosis, thus beginning the formation of a straight line of cells, the *suspensor*. In the meantime, the triploid endosperm cell begins to divide by mitosis to form the *endosperm*, the nutrient material for the embryo. Development continues until the embryo has formed a *radicle*, the rudimentary root; a *plumule*, the rudimentary shoot; and one or two *cotyledons*, seed leaves that contain nutrient materials for the developing embryo. As the seeds are maturing, the fruit develops around them. Eventually the fruit ripens and the seeds are released to begin growth when conditions are correct for that particular species.

MINIQUIZ ON PLANT REPRODUCTION

1. In angiosperms, the seed coat is developed from the

(A) fruit
(B) stigma
(C) sepals
(D) ovule
(E) anthers

2. Pollen is

(A) developed on the anther through the process of meiosis
(B) developed in the stigma and is eventually carried by insects to the anther
(C) the hormone that causes flowering in short-day plants
(D) the structure that contains triploid endosperm
(E) the part of the flower that stimulates cell elongation

3. The structure that connects the stigma to the ovary is the

(A) anther
(B) receptacle
(C) style
(D) sepal
(E) filament

4. The female gametophyte is a structure that would be found in the

(A) endosperm
(B) ovule
(C) fruit
(D) style
(E) anther

5. Double fertilization is the result of

(A) two nuclei in the pollen entering the egg
(B) ethylene inhibition

(C) vegetative reproduction
(D) two pollen grains entering the same anther
(E) phytochrome inhibition in short-day plants

6. The radicle of the plant embryo will develop into the

(A) root
(B) stigma
(C) stem
(D) anther
(E) leaves

7. The primary reproductive cell in the ovule that will go through meiosis is the

(A) microspore mother cell
(B) pollen
(C) megaspore mother cell
(D) zygote
(E) none of these

Answers and Explanations for the Miniquiz

1. **(D)** contains all of the rudimentary cells for seed development including those that form the seed coat. (A) is the structure that develops from the ovary and sometimes includes the receptacle of the flower. (B) is the female part of the flower that has a sticky surface for the reception of pollen that is either wind borne or animal borne. (C) are usually green basal leaves of the flower. (E) are the male reproductive parts of the flower.

2. **(A)** is correct. (B) is a reversed statement. (C) is phytochrome. (D) is the seed. (E) is incorrect because a hormone, not a part, stimulates cell elongation.

3. **(C)** is correct. (A) is part of the stamen. (B) is at the base of the flower and connects the stalk to the sepals, petals, etc. (D) is the green leaf at the base of the flower. (E) is the stalk of the stamen.

4. **(B)** is correct. (A) is the nutrient triploid tissue that develops in the ovule after double fertilization. (C) is the ripened ovary. (D) is the structure that connects the stigma to the ovary. (E) produces the pollen.

5. **(A)** is correct. (D) The anther produces pollen grains. (B, C, and E) are unrelated to the question.

6. **(A)** is correct. (B) is the female flower part that receives pollen from the male flower part. (C) develops from the plumule. (D) develops only during the reproductive phase of the adult plant. (E) develops from the plumule.

7. **(C)** is correct. (A) develops into pollen. (B) is the result of meiosis in microspore mother cells. (D) is the diploid cell that results from the union of two haploid gametes.

ANATOMY AND PHYSIOLOGY OF ANIMALS

THE SKELETAL SYSTEM

Words To Be Defined

HYDROSKELETON	SUTURE	PARATHORMONE
EXOSKELETON	LIGAMENT	CALCITONIN
ENDOSKELETON	SYNOVIAL FLUID	CARTILAGE
AXIAL SKELETON	TENDON	CHONDRICHTHYES
APPENDICULAR SKELETON	RED BONE MARROW	

The earliest evolved support system in animals employs water and is known as the *hydroskeleton.* In animals like hydra, water entrapped in the gastrovascular cavity is used by contractile cells to push against the body wall of hydra. Later, the development of chitin for use as an *exoskeleton,* support tissue outside of the body, and cartilage and bone for use as an *endoskeleton,* support tissue inside of the body, replace the hydroskeleton. Both the exoskeleton and endoskeleton provide a rigid framework for support and to act as a lever system by which selected parts of the body can be moved.

The vertebrate skeleton is composed of two major subdivisions, the *axial skeleton,* the bones that run along the axis of the skeleton, and the *appendicular skeleton,* the bones of the limbs and the girdles to which the limbs are attached. The human skeleton is composed of 206 bones. The appendicular portion is composed of 126 bones while the axial portion is composed of 80 bones. Bones are connected to one another in a variety of ways: some are connected by *sutures,* interlocking "wiggles" of bone that are held tightly together; most are held together by *ligaments.* Joints which require a fair degree of movement are lined by a smooth elastic sheet of connective tissue, and they are lubricated by *synovial fluid.* Skeletal muscles are attached to bones either directly or by way of *tendons.*

Seen under the microscope, bone has many little canals which serve as channels for the blood supply that it receives. In higher vertebrates, bone serves not only as a support tissue but also a tissue in which new blood cells are produced (*red bone marrow*) and in which calcium can be stored for future use. The hormone *parathormone* (secreted by the parathyroid gland) stimulates bone to release calcium ions into the blood while the hormone *calcitonin* (secreted by the thyroid gland) stimulates calcium ions in the blood to be deposited into bone. Calcium ions play an important role in a number of physiological functions.

The other support tissue, *cartilage,* receives no blood supply. Cartilage cells depend on diffusion of nutrients from blood capillaries in the surrounding connective tissues. In the *chondrichthyes,* the sharks and rays, the entire skeleton is composed of cartilage. All early vertebrate skeletons are laid down in a pattern of cartilage which is later replaced by mature bone. Cartilage persists in many areas of flexibility in the higher vertebrate skeleton. These areas are in the external ear, the nose, the larynx, the trachea, the bronchi, the eustachian tube of the middle ear, the intervertebral disks, the breastbone, and the synovial joints.

MINIQUIZ ON THE SKELETAL SYSTEM

1. In the human skeleton, the axial skeleton contains

   (A) 206 bones
   (B) 30 bones
   (C) 80 bones
   (D) 14 bones
   (E) 126 bones

2. Bones do not function in the

   (A) production of blood cells in the higher vertebrates
   (B) support of the vertebrate body
   (C) balance of calcium ions in the bloodstream
   (D) production of calcitonin
   (E) lever systems in the body

3. The exoskeleton of some animals is composed of

   (A) cartilage
   (B) chitin
   (C) bone
   (D) cellulose
   (E) none of these

4. In humans, cartilage is *not* commonly found in the

   (A) external ear
   (B) sutures of the skull
   (C) trachea
   (D) larynx
   (E) intervertebral disks

Answers and Explanations for the Miniquiz

1. **(C)** is correct. (A) is the total number of bones in the human skeleton. (B) is the number of bones in any one of the human arms or legs. (D) happens to be the number of bones in the face. (E) is the number of bones in the appendicular skeleton.

2. **(D)** takes place in the thyroid gland. (A, B, C, and E) are all functions of bone.

3. **(B)** is a mucopolysaccharide that is used in the exoskeleton of insects, crabs, lobsters, etc. (A and C) are support tissues found in endoskeletons. (D) is part of support tissue in plants.

4. **(B)** are often tight, immovable, acartilaginous junctions between bones of the skull. (A, C, D, and E) all contain cartilage of one type or another.

THE MUSCULAR SYSTEM

Words To Be Defined

EFFECTORS
CONTRACTILITY
EXCITABILITY
SMOOTH MUSCLE
SKELETAL MUSCLE
CARDIAC MUSCLE
MUSCLE FIBER
SARCOLEMMA
SARCOMERE
ACTIN
TROPONIN
TROPOMYSIN
MYOSIN
NEUROMUSCULAR JUNCTION
SLOW FIBERS
FAST FIBERS
ANTAGONISM
EXTENSOR
FLEXOR

The muscles of the body of higher animals form a system of *effectors,* cells, tissues, or organs specialized to respond to stimuli provided by the nervous system or the endocrine system. The physiological properties of *contractility,* the ability to shorten, and *excitability,* the ability to transmit stimuli, are particularly well developed in muscle cells. Vertebrates have three kinds of muscle: *smooth muscle,* which is found in the walls of many of the internal organs and blood vessels; *skeletal muscle,* which is associated with and moves the skeletal system; and *cardiac muscle,* which makes up the bulk of the heart.

Skeletal (*voluntary, striated*) muscles are attached to, and move the skeleton. Forty percent of the human body by weight is composed of skeletal muscle. A typical skeletal muscle is composed of numerous *muscle fibers* (cells), each of which traverses the full length of the muscle. A skeletal muscle cell is multinucleate and surrounded by a *sarcolemma,* a membrane analogous to the cell membrane of a uninucleate cell. Each muscle cell is composed of a series of *sarcomeres,* units of myofibrils arranged between adjacent Z-lines (Fig. 3-18). Each sarcomere, in turn, is composed of thin protein filaments (a twisted double strand of *actin* with lesser amounts of *troponin* and *tropomyosin*) intermingled with thick protein filaments (several strands of *myosin* twisted together). The sliding of the actin and myosin filaments results in a shortening of a sarcomere. The shortening of all of the sarcomeres in the muscle causes the muscle contraction. Experimental evidence indicates that cross-bridges between thick and thin protein filaments interact in such a way as to move the thin filaments toward the center of the cell. ATP is needed to provide the energy for the movement of protein fibers, and calcium ions are needed to remove troponin, which serves as a barrier to movement of the thin fibers. ATP is also needed in the process of muscle relaxation. The cross-bridges formed in muscle fiber sliding during contraction need to be broken in order for the muscle to return to the relaxed condition. Each muscle cell is innervated by one to several motor neurons. The small space between the nerve cell ending and the sarcolemma is called the *neuromuscular junction.*

Most skeletal muscle groups are capable of graded responses. In some animals, such as arthropods, there are both excitatory and inhibitory neurons that

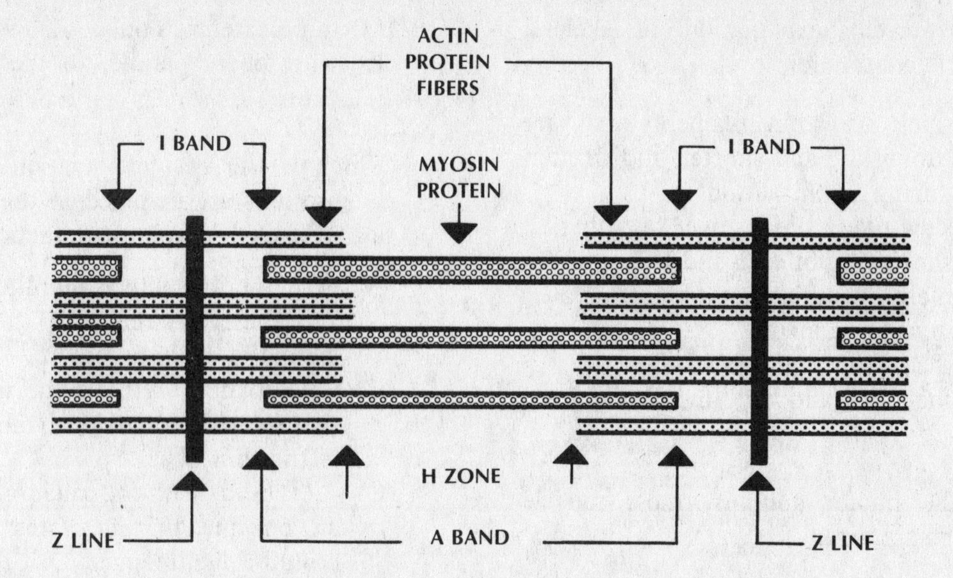

Fig. 3-18 A Typical Sarcomere in Skeletal Muscle

establish reciprocal stimuli in bringing about a graded response. Most invertebrate muscles are composed of *slow fibers,* that is, ones that build up tension slowly. Many vertebrate muscles are also composed of slow fibers. Mammalian muscles have slow fibers but have a greater percentage of *fast fibers*—those that build up tension quickly. The advantage of the fast fibers is in the rapid movement it affords the mammals. Movement is controlled by means of *antagonism,* a muscle arrangement whereby one muscle contracts and at the same time stretches another muscle. When the stretched muscle contracts, it stretches the contracted muscle again. A muscle that causes a joint to straighten is called an *extensor,* while its antagonist which causes a joint to close up is called a *flexor*. The attachment of skeletal muscle to exoskeleton differs from the attachment of skeletal muscle to endoskeletons. In exoskeletons, the muscle generally attaches all along the exoskeleton while in endoskeletons, the muscle generally attaches at a single point of the endoskeleton.

MINIQUIZ ON THE MUSCULAR SYSTEM

1. The microscopic space between the ending of an axon and the sarcolemma is the

   (A) synapse
   (B) terminal bar
   (C) neuromuscular junction
   (D) basement membrane
   (E) pyloric sphincter

2. An enzyme that destroys troponin would most likely affect

   (A) thin filaments of the sarcomere
   (B) thick filaments of the sarcomere
   (C) the neuromuscular junction
   (D) the T-system of muscle
   (E) all lipids

3. The current explanation of skeletal muscle contraction is

   (A) thick and thin filaments that are rubberlike and shorten and thicken during a contraction
   (B) slow fibers sliding over fast fibers
   (C) shortening of each individual sarcomere by sliding actin filaments over myosin filaments
   (D) the spiral screwthread theory
   (E) the "lock and key" mechanism

4. In the higher animals, most muscle groups are arranged

   (A) perpendicular to the direction of pull
   (B) horizontal to the pull of gravity
   (C) in antagonistic pairs
   (D) to be extensors only
   (E) with fibers parallel to the sagittal plane

5. Which is the best explanation of how calcium ions are involved in the mediation of skeletal muscle contraction?

   (A) Calcium ions link to milk protein and thereby strengthen the sarcolemma of each sarcomere.
   (B) Calcium ions remove the troponin that serves as a barrier to movement of the thin fibers.
   (C) ATP binds with the calcium to form an enzyme that activates myosin cross-bridge linkage.
   (D) Calcium is released at the neuromuscular junction where it depolarizes the sarcolemma.
   (E) none of these

Answers and Explanations for the Miniquiz

1. **(C)** is the microscopic junction across which the acetylcholine, released from the axon terminus, travels to the sarcolemma and causes its depolarization. (A) is the microscopic junction between the axon terminus of one neuron and the dendrite or cell body of another neuron. (B) is a structure found where two epithelial cell membranes contact one another. (D) is a structure found at the base of all epithelium. (E) is the muscular sphincter between the pylorus and the duodenum.

2. **(A)** is composed of a twisted double strand of actin with lesser amounts of troponin and tropomysin and would therefore be affected by the enzyme. (B, C, and D) do not relate to the question as they do not involve troponin. (E) Troponin is a protein.

3. **(C)** is correct. (A, B, and D) do not make any sense. (E) is proposed for enzyme-substrate relationships.

4. **(C)** is the way in which the muscles of the body can reestablish themselves once a muscle contracts. (A and B) are incorrect because muscles are arranged in many different directions. (D) is incorrect because muscles bring about different types of movement in the body. (E) is nonsense.

5. **(B)** is correct.

THE DIGESTIVE SYSTEM

Words To Be Defined

AUTOTROPH
HETEROTROPH
HERBIVORE
CARNIVORE
OMNIVORE
DIGESTION
VITAMIN
MINERAL
EXTRACELLULAR
INTRACELLULAR
GASTROVASCULAR CAVITY
MOUTH
ANUS
PHARYNX
CROP
GIZZARD
INTESTINE
MECHANICAL DIGESTION
CHEMICAL DIGESTION
INCISORS
MOLAR
CANINE
SALIVA
AMYLASE
DEXTRINE
MALTOSE
BOLUS
EPIGLOTTIS
TRACHEA
ESOPHAGUS
PERISTALSIS
STOMACH
SMALL INTESTINE
PEPSIN
SPHINCTER
CARDIAC SPHINCTER
PYLORIC SPHINCTER
CHYME
DUODENUM
INTESTINAL JUICE
PANCREATIC JUICE
PANCREATIC DUCT
BILE
GALL BLADDER
COMMON BILE DUCT
JEJUNUM
ILEUM
LARGE INTESTINE
FECES
RECTUM
DEFECATION
ANAL SPHINCTER

There are two major modes of nutrition in life, autotrophic and heterotrophic. *Autotrophs* are organisms that are capable of manufacturing complex foods from simple inorganic molecules from the environment—most plants are autotrophs. *Heterotrophs* are organisms that must ingest preformed complex organic molecules from the environment—animals and fungi are heterotrophs. Animal nutrition breaks down into three subdivisions: *herbivorous,* the consumption of plants; *carnivorous,* the consumption of animals; and *omniverous,* the consumption of both plants and animals. Ingested foods must be *digested,* broken down into simpler molecules, prior to absorption and utilization.

Digestion usually requires the hydrolysis of polymers into monomers, particularly hexose sugars, fatty acids, alcohols, and amino acids. In addition to these relatively large molecules, digestion of complex foods also provides needed *vitamins,* organic compounds that generally function as coenzymes, and *minerals,* inorganic substances that function in numerous physiological mechanisms.

Some digestion is *extracellular,* taking place in a digestive cavity in the body of an animal; some digestion is *intracellular,* taking place in vacuoles within cells. Intracellular digestion is common to protists and a few lower invertebrates, while extracellular digestion predominates among many lower invertebrates and all vertebrates. The first "gut" is found among the Cnidaria. This gut is a blind sac known as the *gastrovascular cavity*. It serves as a cavity for food breakdown and as a mechanism for the circulation of the digested foods. The gastrovascular cavity persists among the Platyhelminthes but becomes a branched structure that increases the surface area for absorption of digested materials. The single opening in the gastrovascular cavity serves both as *mouth*, the structure for the ingestion of food, and *anus*, the structure for the egestion of nondigested materials.

In all of the higher animals, the digestive tract is basically a tube with a mouth at one end and an anus at the other end. During evolution of the higher forms, the simple tubular digestive tract develops into a complex tube with modifications along its length. In annelids, the tube starts with a mouth and is then modified along its length into a muscular *pharynx*, a thin-walled esophagus, a thin-walled *crop*, a muscular *gizzard*, and a straight, muscular *intestine* that terminates in an anus. The highest evolution of the digestive tube is found in the mammals. The mammal digestive tract carries on both *mechanical digestion*, the physical dissolution of food, and *chemical digestion*, the chemical dissolution of food.

The Human Digestive System

The digestive system in human beings has five basic functions: (1) ingestion of food; (2) mechanical and chemical digestion of food; (3) storage and transport of food; (4) absorption of digested foods; (5) formation and evacuation of masses of nondigestable materials and body wastes. Mechanical digestion begins in the mouth with the actions of the teeth. The *incisors* are the chisel-like teeth used for cutting large pieces of food into smaller bits. The *molars* are large, flat teeth used to grind pieces of food into smaller bits. In humans, the *canines* are reduced and generally not particularly functional. In some other mammals, the canines are large and well developed for grasping living prey. In humans, as the food is being mechanically digested, the secretion of the oral cavity, *saliva*, is being added to moisten the mass of food being chewed. Saliva contains the enzyme *amylase* which hydrolyzes starch and glycogen in *dextrines*, small polysaccharides, and *maltose*, a disaccharide. The food does not remain long enough in the oral cavity to say that chemical digestion occurs there. The moistened food is formed into a *bolus*, a ball. The tongue moves the bolus backward toward the *pharynx* where it glides over the *epiglottis*, the structure that acts as a closed "door" preventing food from entering the *trachea*. The bolus passes from the pharynx to the tubular *esophagus*. The food moves down the esophagus by means of *peristalsis*, rhythmic muscular contractions along the tubular portion of the digestive tract. The esophagus ends at the *stomach*, the widest portion of the tract.

The functions of the stomach include: (1) a food-storage structure which controls the time-release of quantities of food into the *small intestine;* (2) further mechanical digestion by the muscular "churning" of food in the stomach; (3) an organ that produces and releases digestive juices that are mixed with the food mass. Stomach digestive juice is composed of acid and digestive enzymes. The main protein-digesting enzyme produced by the stomach wall is *pepsin*. The storage and time-release of food provide a human being with an opportunity to be a discontinuous feeder and thereby spend time carrying on other activities between feedings.

At each end of the stomach there is a circular muscle called a *sphincter*. The one found at the junction between the esophagus and the stomach is called the *cardiac sphincter,* while the one at the junction of the stomach and the small intestine is called the *pyloric sphincter*. Periodic relaxation of the pyloric sphincter permits a squirt of the thick, soupy mixture in the stomach, *chyme,* to move into the first part of the small intestine, the *duodenum.* In the duodenum, more digestive enzymes are added to the chyme. There are digestive enzymes in the *intestinal juice,* produced by the wall of the duodenum, and the *pancreatic juice,* produced by the pancreas and delivered to the duodenum by means of a *pancreatic duct.* In addition, the *bile* produced by the liver, stored in the *gall bladder,* and released into the duodenum through the *common bile duct,* contains salts that aid in the breakdown of fats. The last two portions of the small intestine, the *jejunum* and the *ileum,* absorb the digested foods. The sugars and amino acids pass into the blood vessels in the wall of the small intestine while the partially digested fats pass into the lymphatic vessels in the wall of the small intestine.

The remainder of the material moves along into the *large intestine,* where water and minerals are absorbed into the blood vessels of the wall of the large intestine. Bacteria that normally populate the large intestine produce vitamin K, which is also absorbed into the blood vessels in the wall. The mass of bacteria and nondigested materials form the bulk of material known as *feces*. The feces accumulate in the *rectum* until the time of *defecation,* expulsion of feces. Defecation occurs when there is a relaxation of the *anal sphincter* accompanied by muscular contractions of the wall of the rectum.

Digested foods are absorbed into the bloodstream and lymphatic system by means of both passive and active transport mechanisms. The structure of the small intestine provides an enormous surface area for absorption. Although the liver will be discussed with the circulatory system, we should mention here that the liver plays a major role in controlling the fate of the absorbed sugars, amino acids, vitamins, and minerals. The liver stores sugars as glycogen, converts excess sugars into fat, deaminates amino acids, manufactures blood proteins, and produces nitrogen wastes.

MINIQUIZ ON THE DIGESTIVE SYSTEM

1. Three organs that contribute to chemical digestion but are *not* part of the digestive tract proper are the

   (A) spleen, liver, and salivary glands
   (B) stomach, liver, and esophagus
   (C) salivary glands, stomach, and teeth
   (D) liver, salivary glands, and pancreas
   (E) malleus, incus, and stapes

2. Chemical digestion in humans is usually completed in the

   (A) mouth
   (B) large intestine
   (C) stomach
   (D) small intestine
   (E) pharynx

3. Which function is *not* normally ascribed to the liver?

   (A) destruction of old red blood cells
   (B) production of enzymes that assist the digestive system
   (C) detoxification of the bloodstream
   (D) production of plasma proteins for the circulatory system
   (E) production of red blood cells in adults

4. The human teeth used to grind and crush are the

   (A) incisors
   (B) canines
   (C) molars
   (D) gizzards
   (E) eyeteeth

5. Mechanical digestion occurs in the

   (A) stomach
   (B) crop
   (C) esophagus
   (D) duodenum
   (E) jejunum

6. The first digestive system appears to have evolved among the

   (A) vertebrates
   (B) cnidarians
   (C) annelids
   (D) platyhelminthes
   (E) protozoa

7. The *major* advantage of the mouth-to-anus digestive tract is that it permits

   (A) the ingestion of pieces of food much larger than the gastrovascular cavity
   (B) animals without teeth a means for mechanical digestion
   (C) the animal to be a discontinuous feeder
   (D) the animal to expel nondigested materials
   (E) continuous feeding

For questions 8–11 use the following key

(A) mouth
(B) large intestine
(C) small intestine
(D) stomach
(E) none of these

8. The structure that supports the bacteria which produce vitamin K.

9. The structure that contributes amylases to the digestive juices.

10. The structure that carries out mechanical digestion and produces pepsin.

11. The structure that produces bile.

Answers and Explanations for the Miniquiz

1. **(D)** all contribute chemicals that participate in digestion. The spleen is an organ of the circulatory system. The teeth and esophagus do not contribute any chemicals that participate in digestion. (E) are the three ear bones.

2. **(D)** is the organ in which chemical digestion is completed. (A and C) are both organs in which chemical digestion occurs but is not completed. (B and E) do not produce any chemicals that contribute to digestion.

3. **(B)** is the only function not ascribed to the liver. The liver produces bile which is composed of salts and pigments but no digestive enzymes.

4. **(C)** are the grinding teeth. (A) are used for cutting. (B) are for tearing and holding. (D) are used for grinding but are not teeth. (E) is another term for the canines.

5. **(A)** carries out both mechanical and chemical digestion. (B) is an organ of storage. (C) is an

organ that conducts a food bolus from the pharynx to the stomach. (D) is the first portion of the small intestine and contributes only to chemical digestion. (E) is the last portion of the small intestine and is an area of digested food absorption.

6. **(B)** are the simplest organisms to exhibit any type of "digestion system," the gastrovascular cavity. (A and C) both possess a mouth-to-anus digestive system which is the second type of system to evolve. (D) have a more complex, branched, gastrovascular cavity than do the Cnidaria. (E) Systems require complex multicellularity. Protozoa are one-celled animals.

7. **(C)** allows animals to spend time on an assortment of activities in addition to feeding.

8. **(B)** is the part of the tract that supports the growth of a bacterial symbiont. This symbiont produces vitamin K that is absorbed into the human bloodstream.

9. **(A)** contains the salivary glands, which produce saliva. Saliva contains the enzymes known as amylases.

10. **(D)** does both. (A) carries out both mechanical and chemical digestion but does not produce pepsin.

11. **(E)** is correct as it is the liver that produces bile.

THE CIRCULATORY SYSTEM

Words To Be Defined

VASCULAR SYSTEM
CIRCULATORY SYSTEM
BLOOD
HEART
CAPILLARY
ARTERY
VEIN
LYMPH
COELOM
CLOSED BLOOD CIRCULATION
OPEN BLOOD CIRCULATION
HEMOLYMPH
HEMOCOEL
TRACHEAL SYSTEM
SINGLE CIRCULATION
DOUBLE CIRCULATION
PULMONARY CIRCULATION
SYSTEMIC CIRCULATION
ATRIUM
VENTRICLE
SINUS VENOSUS
CONUS ARTERIOSUS
VENTRAL AORTA
DORSAL AORTA

Types of Transport Systems

The transport of materials around the body must occur even in the simplest multicellular forms. Animals have evolved a variety of *vascular systems,* fluid systems that enable the transport of gases, digested foods, wastes, etc. A vascular system that uses muscular organs to propel fluids through vessels is known as a *circulatory system.* In mammals, the circulatory system consists of a *blood*-conducting muscular pump called the *heart,* blood vessels called *capillaries, arteries,* and *veins,* and a *lymph*-conducting set of capillaries and veins. The reason that there are no lymph arteries will be discussed later. In addition to gas exchange, food transport, and waste elimination, the circulatory system of higher vertebrates transports a variety of defense cells, defense molecules in the forms of antigens, control molecules in the form of hormones, and heat.

The simplest circulatory system, the saclike gastrovascular cavity, is found in the Cnidaria. The food that is partially broken down by extracellular digestion is circulated around to the cells lining the gastrovascular cavity. These cells phagocytize the food particles and complete digestion intracellularly. In a cnidarian like *Hydra*, the gastrovascular cavity is a simple sac, while in the starfish *Aurelia*, the gastrovascular cavity elaborates into a branched structure with greater surface area for absorption. In the free-living platyhelminthes like *Planaria*, the gastrovascular cavity becomes very highly branched and rebranched. With the evolution of the *coelom*, a body cavity in which internal organs are suspended in a membrane but have a degree of movement, there developed a *closed blood circulation*, a system of vessels which prevents the blood from directly mixing with the cells. The annelids exhibit one of the simplest closed circulations. Earthworms possess five pairs of "hearts" that pump blood. Arteries transport blood away from the hearts while veins transport blood to the hearts. Capillaries connect arteries and veins. All exchanges between the blood and any part of the body or the external environment take place in the capillaries. As the division of labor increases in evolution, the physiology of an organism becomes more efficient. The development of the circulatory system is one example of the specialization of function.

A number of invertebrates have developed an *open blood circulation*, a system of a few, disconnected blood vessels between which are sinus areas that allow the blood to intermingle with other body fluids and the body cells themselves. There is a main muscular pump that moves the blood in one general direction with auxillary pumps at the bases of various appendages. The blood in insects, e.g., *hemolymph*, permeates the entire body cavity, the *hemocoel*. The gas exchange afforded by the circulation of hemolymph is insufficient to meet the needs of the animal and therefore insects have also evolved a complex, branched tubular gas exchange system called the *tracheal system*.

The entire group of vertebrates has evolved closed circulatory mechanisms. In the fishes, there is a *single circulation*, blood moving once through the heart before completing a circuit through the body. The Amphibia, Reptilia, Aves, and Mammalia all possess a *double circulation*, blood moving twice through the heart before completing a circuit through the body. Double circulation includes the *pulmonary circuit*, the movement of blood through the right side of the heart to the lungs and back to the left side of the heart, and the *systemic circuit*, the movement of the blood through the left side of the heart to the body and back to the right side of the heart. In amphibians and reptiles the heart is not fully divided into a right and left side and therefore blood that is rich in oxygen is diluted by blood that is poor in oxygen. In the birds and mammals the left and right sides of the heart are fully divided and therefore the blood, rich in oxygen, is not diluted.

The circulatory systems of vertebrates exhibit two evolutionary trends. The first is the increase in oxygenated blood in the dorsal aorta and the second is the increase in blood pressure in the dorsal aorta. In fishes, the heart consists of a two-chambered structure, the *atrium* and the *ventricle*, with a preheart *sinus venosus*, and a postheart *conus arteriosus*. From the fish heart blood moves anteriorly through a short *ventral aorta* to the network of vessels in the gills. From the gills, the blood moves into a *dorsal aorta* that then distributes blood anteriorly and posteriorly to all parts of the body. Because of the diminished flow of blood through the narrow confines of the gills, the pressure of the blood in the dorsal aorta is significantly lower than the pressure exerted by the heart on the blood. Double circulation allows oxygenated blood to be returned to the heart for an additional push to all parts of the body. It is for this reason that the blood pressure in the dorsal aorta of the higher vertebrates is higher than that found in fishes.

Amphibians and reptiles have a three-chambered heart, left and right atrium, and a single ventricle that may or may not be partially partitioned into left and right sides. Birds and mammals have true four-chambered hearts. Blood from the body passes through the right atrium and right ventricle and moves to the lungs while blood from the lungs passes through the left atrium and left ventricle back to the body.

Mammal Blood and Lymph

Words To Be Defined

BLOOD
PLASMA
FORMED ELEMENT
ERYTHROCYTE
LEUKOCYTE
THROMBOCYTE
PLASMA PROTEIN
SERUM
ERYTHROPOIETIN
HEMOGLOBIN
ANEMIA
THROMBOPLASTIN
PROTHROMBIN
FIBRINOGEN
FIBRIN
BLOOD CLOT
INTERSTITIAL FLUID
LYMPHATIC SYSTEM
LYMPH CAPILLARY
LYMPH
LYMPH VEIN
THORACIC DUCT
RIGHT LYMPH DUCT
B-LYMPHOCYTE
T-LYMPHOCYTE
ANTIBODY
ANTIGEN
GLYCOPROTEIN
AGGLUTINATION
RH

Blood Composition

Blood is a type of connective tissue with a liquid matrix. Roughly 55% of the volume of blood is matrix (*plasma*) and 45% is composed of blood cells (*formed elements*). There are three general types of blood cell: the *erythrocytes*, red blood cells; the *leukocytes*, white blood cells; and the *thrombocytes*, platelets.

Plasma is composed of about 90% water, 7–8% proteins (*plasma proteins* which are involved in blood clotting), and about 3% substances that vary, i.e., gases, wastes, foods, hormones, and ions. Plasma with the plasma proteins removed is known as *serum*.

The composition of blood is regulated by a number of mechanisms. The kidney regulates the salts, water, and nitrogen wastes in the blood, and under conditions of reduced oxygen supply due to reduced numbers of erythrocytes produces a hormone called *erythropoietin*. This hormone stimulates bone marrow to increase its production of erythrocytes. As the number of erythrocytes in the blood increases to normal, the increased oxygen supply to the kidney diminishes the production of the hormone. The composition of blood is also regulated in a number of ways by the liver, which regulates blood glucose, amino acids, plasma proteins, toxins, etc. It also destroys old erythrocytes and conserves the iron atoms while the remainder of the hemoglobin contributes to the formation of bile.

The erythrocytes are the most numerous blood cells, and at sea level vary from 4.5–5.5 million per cubic millimeter of blood in humans. Erythrocytes in the adult are produced in the bone marrow of the long bones. Immature erythrocytes have nuclei which disintegrate as the erythrocytes reach maturity. Mature erythrocytes are essentially "bags" of *hemoglobin*, the red protein that is capable of loosely binding with oxygen. *Anemia* is the condition in which there is a lower than normal number of erythrocytes or a lower than normal amount of hemoglobin.

The leukocytes are less numerous than erythrocytes, normally varying from 8,000–9,000 per cubic millimeter of blood in humans. Leukocytes in the adult are produced in the bone marrow of long bones and in the lymphoid tissues of the body. Leukocytes are nucleated blood cells that participate in defense against disease.

The thrombocytes are in reality cell fragments that contribute to the blood-clotting mechanism. In the normal human adult there are approximately 200,000 per cubic millimeter of blood. Thrombocytes contain an enzyme called *thromboplastin* which catalyzes the first change that leads to the clotting of blood. Thromboplastin, in the presence of calcium ion, facilitates the conversion of the plasma protein *prothrombin* into *thrombin*, which then catalyzes the conversion of the plasma protein *fibrinogen* into *fibrin*. It is the fibrin that forms the meshwork around which a *blood clot* develops. The fibrin meshwork traps cells and debris sealing off the injured part.

The Lymphatic System

Leakage from the blood circulation serves the very important role of bathing the living cells of a tissue. This blood-derived fluid is called *interstitial fluid*. The *lymphatic system* is one of the two mechanisms by which excess tissue fluids are drained from the tissues. The other functions of the lymphatic system include carrying large molecules such as hormones, fats, and proteins, temporarily storing fluids taken into the body, and producing leukocytes concerned with the protection of the body. *Lymph capillaries* are the vessels that collect the *lymph* and conduct it to the *lymph veins*. The lymph veins eventually come together to form the *thoracic duct* and the *right lymph duct*. These vessels empty into the venous portion of the blood circulation at a point near the heart. The lymph nodes play an important role in the immune system of the body. Lymphocytes are either produced in the bone marrow (*B lymphocytes*), or are produced in bone marrow but undergo additional development in the thymus glands (*T-lymphocytes*). B-lymphocytes produce circulating *antibodies,* proteins that combat foreign proteins called *antigens*. Each T-lymphocyte produces a unique cellular antibody that attacks a specific antigen.

Blood Groups

The two important blood groups in human beings are ABO and Rh. The ABO blood group is the result of the interaction of three genes. The genes for A and B are codominant while each is dominant to the gene for O. Each of the codominant genes confer upon erythrocyte membranes a specific *glycoprotein,* a protein bound to a carbohydrate. Individuals who have blood type A have A antigens on their erthyrocyte membranes and B antibodies in their plasma, while those with blood type B have B antigens on their erythrocytes and A antibodies in their plasma. Individuals who have blood type O have no antigens on their erythrocytes but do have both A antibody and B antibody in their plasma. Individuals who have blood type AB have both A and B antigens on their erythrocytes but no antibody in their plasma. Improper mixing of blood during a transfusion causes an interaction of antigen and antibody that results in erythrocytes sticking together, a process called *agglutination.*

Rh blood factors are independent of the ABO blood factors. ABO antibodies are always present in the plasma while Rh antibodies are produced only when Rh antigens are in the blood. Individuals who are Rh+ have the Rh antigens on the membranes of their erythrocytes while those who are Rh− do not. If a transfusion between Rh incompatible individuals occurs, the Rh− recipient of Rh+ blood will begin to manufacture Rh antibodies. It usually takes 2–4 months for the sensitization to take place. If the sensitized individual receives another transfusion of Rh+ blood, the erythrocytes agglutinate.

Circulation of Blood Through the Heart

Words To Be Defined

SUPERIOR VENA CAVA	PULMONARY ARTERY	AORTA
INFERIOR VENA CAVA	PULMONARY VEIN	SA NODE
RIGHT ATRIUM	LEFT ATRIUM	AV NODE
TRICUSPID VALVE	BICUSPID VALVE	SYSTOLE
RIGHT VENTRICLE	LEFT VENTRICLE	DIASTOLE

In spite of the single structure, we can think of the heart as being composed of a “right heart” and a “left heart” (see Fig. 3-19). In humans, the blood from around the body enters the right side of the heart from both the *superior vena cava,* the vessel that collects all of the blood from above the heart, and the *inferior*

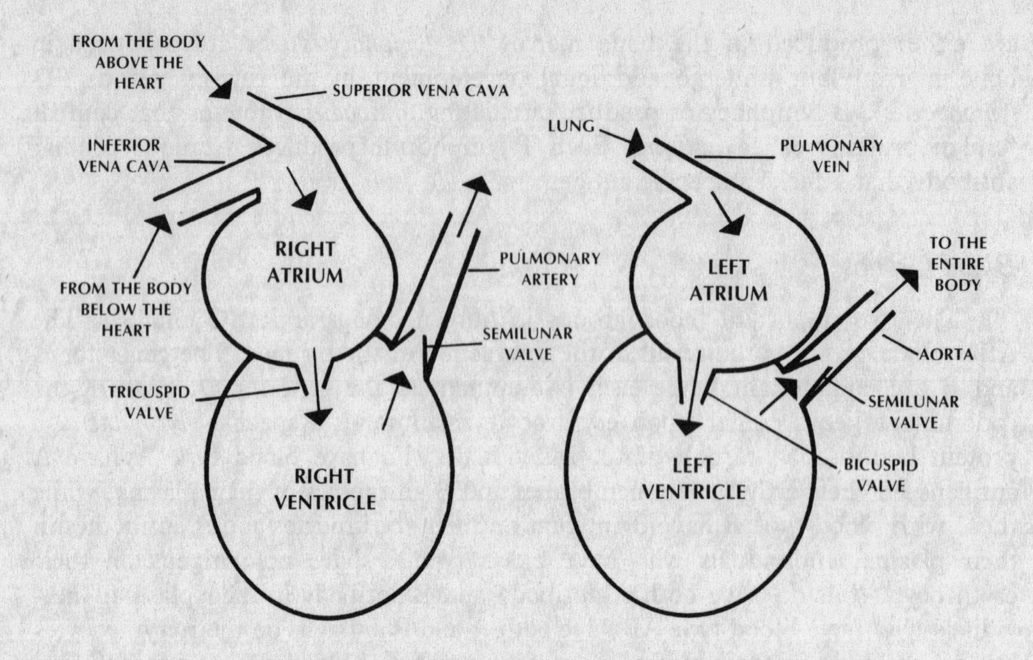

Fig. 3-19 Circulation of Blood Through the Human Heart

vena cava, the blood vessel that collects all of the blood from below the heart. All of the blood enters the first chamber of the heart, the *right atrium* and then moves through the *tricuspid valve* and into the second chamber, the *right ventricle*, and finally through a *semilunar valve* into the *pulmonary artery*. The blood enters the lung where gas exchange between the blood and the external environment occurs. The blood continues to the *pulmonary veins* that lead it to the third chamber of the heart, the *left atrium*. Blood from the left atrium moves through the *bicuspid valve (mitral valve)* into the fourth chamber, the *left ventricle*, and on out through another semilunar valve and into the *aorta*, the large artery that distributes blood to the body. The aorta is the origin of a number of blood vessel branches that distribute blood to the head, arms, legs, and trunk region.

The mechanism of heartbeat begins in the heart in a mass of tissue, the *sinoatrial* (SA) *node* (pacemaker) which is located in the wall of the right atrium. When stimulated, the SA node initiates the contraction of the atria and also stimulates the atrioventricular (AV) *node*. The AV node in turn initiates the contraction of the ventricles. The contraction of the ventricles is known as *systole* while the relaxation of the heart is known as *diastole*. The sounds associated with the heartbeat have been described as "lubb-dubb." The "lubb" comes from the ventricles contracting while the "dubb" is the result of the snapping shut of the semilunar valves.

MINIQUIZ ON THE CIRCULATORY SYSTEM

1. Open circulation is characteristic of
   (A) annelids
   (B) insects
   (C) vertebrates
   (D) mammals
   (E) birds

2. The complete four-chambered heart is characteristic of all
   (A) mammals and birds
   (B) reptiles
   (C) vertebrates
   (D) fishes
   (E) sharks and rays

3. The main component of blood plasma is
   (A) plasma protein
   (B) hormone
   (C) water
   (D) wastes and food
   (E) gas

4. Veins can be said to be blood vessels that
   (A) always carry blood rich in carbon dioxide and poor in oxygen
   (B) carry blood away from the heart and blood to the heart
   (C) are important because they allow materials to move in and out of blood
   (D) are only found in the arms and legs of the higher vertebrates
   (E) are not found in the walls of the heart

Use the following key to answer questions 5–8

(A) T-lymphocyte
(B) antigen
(C) antibody
(D) erythrocyte
(E) none of these

5. The blood cell that undergoes modification in the thymus gland and has a nucleus throughout its "life."

6. The protein that is found on the membranes of erythrocytes

7. The protein found in blood plasma

8. The protein that creates the meshwork that is the foundation of a blood clot.

9. Agglutination of blood is the result of the
   (A) antibody in plasma combining with erythrocyte membrane antigen and causing the erythrocytes to stick to one another
   (B) formation of a meshwork of fibrin that traps blood cells and debris
   (C) Rh+ blood mixing with A blood type
   (D) T-lymphocytes releasing antigens into the plasma and attaching antibodies to platelets
   (E) none of these

10. The mitral valve is found between the
   (A) right atrium and right ventricle
   (B) right ventricle and the pulmonary artery
   (C) left atrium and the left ventricle
   (D) left ventricle and the aorta
   (E) inferior vena cava and right atrium

11. Blood passes through the heart twice without mixing during those passages in
   (A) amphibians
   (B) fishes
   (C) reptiles
   (D) annelids
   (E) insects

12. Hemolymph is typical of
   (A) fishes
   (B) reptiles
   (C) mammals
   (D) insects
   (E) birds

Answers and Explanations for the Miniquiz

1. **(B)** typically have an open circulation. (A, C, D, and E) all have closed circulation.

2. **(A)** typically have a four-chambered heart. (B) typically have a three-chambered heart. (C) include the fishes, reptiles, birds, and mammals. (D and E) have two-chambered hearts.

3. **(C)** represents 90% of plasma. (A) represents 7% of plasma. (B, D, and E) are a small part of the 3% variables in the plasma.

4. **(E)** is correct. Veins are blood vessels that carry blood to the heart only. (A) is incorrect because the pulmonary vein carries oxygen-rich blood from the lung back to the left side of the heart. (B, D, and E) are not correct. (C) is correct for capillaries, not veins.

5. **(A)**

6. **(B)**

7. **(C)**

8. **(E)** The protein that forms the meshwork in a blood clot is fibrin.

9. **(A)** is correct. (B) is the mechanism that begins in bloodclot formation. (C) Rh and ABO are independent blood factors. (D) is pure nonsense.

10. **(C)** is correct. (A) is separated by the tricuspid valve. (B and D) are each separated by semilunar valves. (E) There is no valve at this location.

11. **(E)** is correct. Only birds and mammals have true four-chambered hearts. (A and C) have three-chambered hearts. (B) have two-chambered hearts. (D and E) have a heart arrangement very different from the heart in vertebrates.

12. **(D)** have a circulatory fluid called hemolymph which bathes the tissues directly. (A, B, C, and E) all possess a circulatory fluid called blood which is separated from the tissues by a closed system of vessels.

THE EXCRETORY SYSTEM

Words To Be Defined

FLAME CELL
MALPIGHIAN TUBULE
NEPHRIDIUM
KIDNEY
NEPHRONS
RENAL TUBULE
GLOMERULUS
BOWMAN'S CAPSULE
RENAL CORPUSCLE
PROXIMAL CONVOLUTED TUBULE
LOOP OF HENLE
DISTAL CONVOLUTED TUBULE
COUNTERCURRENT MECHANISM
RENAL CORTEX
RENAL MEDULLA
RENAL PELVIS
URINE
URETER
URINARY BLADDER
URETHRA
MICTURITION
DEAMINATION
URIC ACID
CLOACA
UREA
ALDOSTERONE
ANGIOTENSIN
ANTIDIURETIC HORMONE
RENIN
ANGIOTENSINOGEN

The maintenance of a balanced chemical composition in the body is important. The intake and utilization of environmental substances results in the production of a variety of wastes which need to be eliminated. Small, simple organisms use diffusion to excrete wastes into the environment. Any mechanism of excretion must enable the separation of materials useful to the organism from those materials that are nontoxic and toxic wastes; e.g., water, carbon dioxide, ammonia, and salts. Most of the invertebrates have some type of excretory system that employs a set of tubes that collect and convey wastes to the outside of the body.

The simplest excretory system is found among the platyhelminthes. Specialized cells, *flame cells,* accumulate body wastes and release them in collecting ducts along each side of the body. These ducts conduct the excretions to one of two posterior openings to the environment. Most arthropods possess a complex set of collecting structures and conducting tubules, the *Malpighian tubules.* The annelids utilize a complex tubule structure known as the *nephridium.* Pairs of nephridia are found in each segment of the body. The fluid from the segment immediately anterior to the paired nephridia is filtered through these nephridia. Among the vertebrates, the excretory system employs some form of a *kidney.* The main function of the kidney is to regulate the concentration of the wastes in blood. The kidney carries out ultrafiltration of blood, the separation of plasma (filtrate) from the cellular components of blood, reabsorption from the filtrate, the return of useful substances to the bloodstream, and secretion into the filtrate; that is, the deposit of certain wastes directly into the developed filtrate.

The kidney of higher vertebrates is generally a collection of thousands of functional units called *nephrons.* Each nephron is a unit composed of a capillary network around a *renal tubule.* The capillary network begins as a tuft of capillaries, the *glomerulus,* within an expanded, hollow cuplike portion of the tubule, the *Bowman's capsule.* The combination of the glomerulus and the Bowman's capsule is referred to as the *renal capsule.* Blood from the renal artery enters glomeruli where much of the plasma and its components are filtered from the blood. The filtrate moves from the capsule and next passes through the *proximal convoluted tubule,* the *loop of Henle,* the *distal convoluted tubule,* and moves on to a collecting duct. As the filtrate passes through the different portions of the tubule, its composition is changed by reabsorption and secretion in various segments along the way. The blood that remains in the circulatory system moves from the glomerulus into an arteriole and then into a second bed of capillaries that surrounds the loop of Henle. The descending and ascending portions of the loop of Henle provide a *countercurrent mechanism,* that allows for the constant movement of ions and small molecules, permitting this region of the kidney to maintain a relatively low and uniform osmotic potential. The outer region of the kidney, the *renal cortex* (Fig. 3-20), contains all of the renal corpuscles and convoluted tubules. The inner region of the kidney, the *renal medulla,* contains the loops of Henle and the capillaries that surround them.

All of the collecting ducts empty their contents into the *renal pelvis,* a large hollow area in the kidney. Collected concentrated waste, *urine,* is then transported through a thin-walled tube, the *ureter,* to the saclike *urinary bladder.* The urinary bladder is capable of moderate distention and can collect a fairly large quantity of urine before it requires emptying. As the urinary bladder distends, it begins to stimulate messages to the brain. When the organism contracts the muscles of the urinary bladder, the urine is forced out of the body through the tubular *urethra.* The expulsion of urine from the body is known as *micturition.*

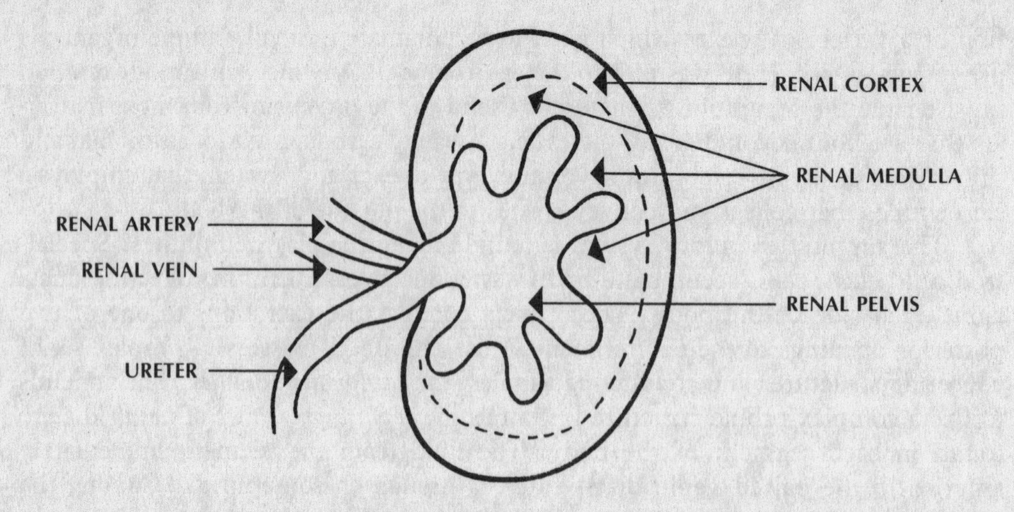

Fig. 3-20 Basic Structure of the Kidney

The *deamination* of amino acids, removal of the amine group, is the most common source of nitrogen waste in animals. The amine groups are turned into ammonia, a very toxic compound. Many aquatic and some land animals excrete ammonia but most land vertebrates must circulate nitrogen wastes for a period of time and therefore cannot tolerate nitrogen in the form of ammonia. Insects, reptiles, and birds convert ammonia to *uric acid*. Reptiles and birds excrete a dilute uric acid solution into a *cloaca*, a compartment common to the digestive, reproductive, and excretory systems. Mammals and other land forms convert ammonia to *urea*, a less toxic compound than ammonia.

The kidney is a complex organ that is under a number of controls. The formation and composition of urine are controlled by several hormones. Among these are *antidiuretic hormone* (ADH), the posterior pituitary protein hormone that increases the body retention of water in the formation of urine; *aldosterone*, the adrenal cortex steroid hormone that promotes the increased reabsorption of sodium from the filtrate in the ascending portion of the loop of Henle; *angiotensin*, the hormone that causes vasoconstriction and an increase in aldosterone production; and *renin*, a protein hormone that functions as an enzyme which promotes the conversion of *angiotensinogen*, a blood protein, into angiotensin.

MINIQUIZ ON THE EXCRETORY SYSTEM

1. Urine from the renal pelvis moves into the urinary bladder through the

   (A) ureter
   (B) collecting duct
   (C) urethra
   (D) renal artery
   (E) none of these

2. The first structure to be modified for the accumulation and elimination of excretory materials is the

   (A) nephridium
   (B) nephron
   (C) flame cell
   (D) Malpighian tubule
   (E) glomerulus

3. The hormone that promotes the increased reabsorption of sodium from the filtrate in the ascending portion of the loop of Henle is

    (A) antidiuretic hormone
    (B) aldosterone
    (C) angiotensin
    (D) renin
    (E) rennin

4. The main excretory material in reptiles and mammals is

    (A) urea
    (B) ammonia
    (C) uric acid
    (D) ADH
    (E) LH

Use the following key to match the structure with the functions specified in questions 5–8.

(A) renal corpuscle
(B) ureter
(C) renal pelvis
(D) urinary bladder
(E) renal cortex

5. Immediately collects the urine from all of the collecting ducts.

6. The location of the renal corpuscles.

7. The organ that is part of the excretory system.

8. The structure in which the filtration of blood takes place.

Answers and Explanations for the Miniquiz

1. **(A)** is correct. (B) is part of the complex tubule arrangement in the kidney. (C) is the tube that conducts urine from the urinary bladder to the outside of the body. (D) is a blood vessel.

2. **(C)** is correct as it is the first excretory structure, found in platyhelminthes. (A) is a complex, modified excretory tubule found in annelids. (B) is one of thousands of similar functional units of the kidney. (D) is the excretory structure found in many arthropods. (E) is part of the nephron.

3. **(B)** is correct. (A) is the hormone that promotes reabsorption of water from forming urine. (C) is the hormone that causes vasoconstriction and promotes aldosterone secretion by the adrenal cortex. (D) is the hormone that acts as an enzyme in the conversion of angiotensinogen to angiotensin. (E) is a gastric enzyme that digests milk protein.

4. **(A)** is correct. (B) is the common nitrogen excretory material in many aquatic invertebrates and vertebrates. (C) is the common nitrogen excretory material in some arthropods, most amphibians, and most birds. (D and E) are hormones.

5. **(C)** is correct.

6. **(E)** is correct

7. **(D)** is the only organ listed.

8. **(A)** is correct.

THE NERVOUS SYSTEM

Words To Be Defined

NEURON
SOMA
DENDRITE
AXON
UNIPOLAR NEURON
BIPOLAR NEURON
MULTIPOLAR NEURON
SENSORY NEURON
CNS
EFFECTOR
INTERNEURON
MOTOR NEURON
SYNAPSE

GLIAL CELL
SCHWANN CELL
MYELIN SHEATH
NODE OF RANVIER
RESTING POTENTIAL
SODIUM–POTASSIUM PUMP
DEPOLARIZATION
HYPERPOLARIZATION
EXCITATION
THRESHOLD
ACTION POTENTIAL (IMPULSE)
SUMMATION

TEMPORAL SUMMATION
SPATIAL SUMMATION
REFRACTORY PERIOD (RP)
ABSOLUTE RP
RELATIVE RP
SYNAPTIC CLEFT
PRESYNAPTIC MEMBRANE
POSTSYNAPTIC MEMBRANE
NEUROTRANSMITTER
SYNAPTIC BULB
ACETYLCHOLINE
NOREPINEPHRINE

Basic Structure of Nervous Tissues

Regulation and coordination of physiological activities and structural elements is the function of both the nervous system, a generally rapid-response system, and the endocrine system, a generally slow-response system. The functional unit of the nervous system is the *neuron.* It is generally composed of a large cell body, *soma,* and two cell processes, the *dendrites* (a process that conveys information to the soma) and an *axon* (a process that conveys information away from the soma). *Structurally,* there are three general types of neurons (Fig. 3-21): the *unipolar neuron* has one main process that bypasses the soma; the *bipolar neuron* has one dendrite on one side of the soma and one axon on the other side of the soma; and the *multipolar neuron* has numerous dendrites and a single axon. *Functionally,* there are three general types of neurons: the *sensory neuron* carries information from sense receptors to the *central nervous system* (CNS), the *motor neuron* carries information from the CNS to *effectors* (muscles or glands), and the *interneuron* connects a sensory neuron with a *motor neuron.* Information moving from one neuron to another has to cross a small open space, the *synapse.*

Other types of cells in the nervous system are the *glial cells,* several types of specialized connective tissue cells that support neurons and scavenge the neuron environment, and the *Schwann cells,* specialized connective tissue cells that form and wrap a fatty layer around the axon of some neurons. This fatty layer, the *myelin sheath,* serves as an insulation for the axon. The spaces formed by the

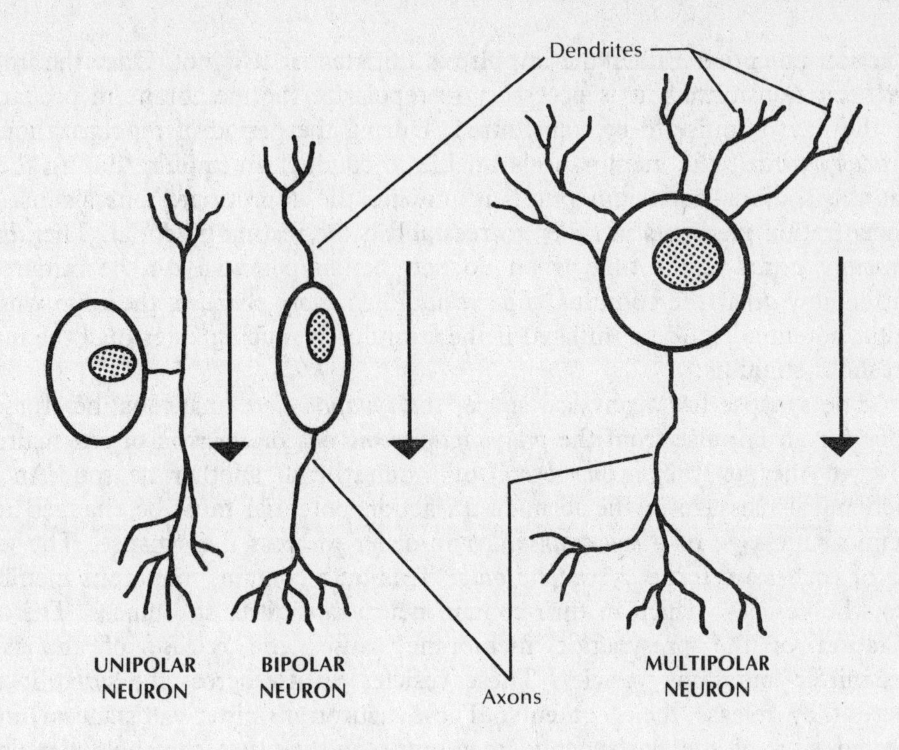

Fig. 3-21 Structural Types of Neurons

junction of adjacent Schwann cells are known as the *nodes of Ranvier*. These nodes are regions of exposed axon.

An asymmetric distribution of ions across the neuron membrane provides the membrane with a *resting potential,* a difference in potential across a membrane. The resting potential is maintained by a *sodium–potassium pump* (Na–K pump), a biochemical mechanism that expells Na+ from the cells while accumulating K+ from the environment. These movements are against concentration gradients and therefore require the expenditure of energy in the form of ATP. Some stimuli can cause a local membrane *depolarization,* a decrease in the potential across the membrane. Some stimuli can cause *hyperpolarization,* an increase in the potential across the membrane. If depolarization or *excitation* of a membrane reaches *threshold,* a level of depolarization that must be exceeded in order to create a wave of depolarization, then an *impulse* or *action potential* is transmitted. Below threshold, stimuli do not generate an impulse or action potential. Hyperpolarization of the membrane also serves as an inhibitor of impulse initiation.

As there is cytoplasmic resistance, there is the possibility for weak threshold stimuli to dissipate during impulse transmission. The phenomenon of *summation,* the addition of numerous dendrite and soma stimuli, prevents the dissipation of the information being conducted by an impulse. *Temporal summation* is dependent upon the time interval between stimuli while *spatial summation* is dependent upon the space overlap of two stimuli on the surface of some part of the neuron. Stimuli above the threshold invoke an "all or none" response; that is, there are no degrees of action potential. Either the impulse is initiated or it is not. Once the impulse has been transmitted, it is necessary to repolarize the membrane in preparation for the next impulse to be transmitted. During the period of repolarization, the *refractory period*, the membrane is unable to conduct an impulse due to the fact that the sodium–potassium pump is moving the appropriate ions against their concentration gradients in order to reestablish the resting potential. The *absolute refractory period* is the time when no new action potential can be initiated no matter how great the stimulus. The *relative refractory period* is the time when an action potential could be initiated if the stimulus is much greater than the normal threshold stimulus.

The synapse has a physical space, the *synaptic cleft,* that must be crossed in order for an impulse from the *presynaptic membrane* of an axon of one neuron to move to the *postsynaptic membrane* of a dendrite of another neuron. An electrochemical message in the form of an action potential must be changed into a chemical message of a *neurotransmitter* in order to cross the synapse. The terminus of each axon forms a *synaptic bulb*. This bulb contains numerous membrane sacs, the vesicles, which in turn contain neurotransmitter substances. The depolarization of the presynaptic membrane causes the release of the neurotransmitter-containing vesicles. These vesicles diffuse across the synaptic cleft, where they release their contents. These neurotransmitter substances bind to selected sites of the postsynaptic membrane and cause a threshold depolarization. The most common neurotransmitter substances are *acetylcholine* and *noradrenalin* (*norepinephrine*).

Organization of the Nervous System

Words To Be Defined

NERVE NET
GANGLIA
PNS
MENINGES
NERVE
FOREBRAIN
MIDBRAIN
HINDBRAIN
DIENCEPHALON
THALAMUS
HYPOTHALAMUS
POSTERIOR PITUITARY
TELENCEPHALON
OPTIC LOBE
CEREBRUM
PONS
MEDULLA OBLONGATA
SOMATIC NERVOUS SYSTEM
AUTONOMIC NERVOUS SYSTEM
SYMPATHETIC SYSTEM
PARASYMPATHETIC SYSTEM
CRANIAL NERVE
SPINAL NERVE
PHOTORECEPTOR
CHEMORECEPTOR
MECHANORECEPTOR
THERMORECEPTOR
PROPIOCEPTOR
STATOCYST
HAIR CELL
EAR
PINNA
EXTERNAL AUDITORY MEATUS
TYMPANIC MEMBRANE
MALLEUS
INCUS
STAPES
OVAL WINDOW
EUSTACHIAN TUBE
SEMICIRCULAR CANAL
SACCULE
UTRICLE
COCHLEA
OTOLITH
VESTIBULAR CANAL
TYMPANIC CANAL
PERILYMPH
COCHLEAR DUCT
ENDOLYMPH
ORGAN OF CORTI
BASILAR MEMBRANE
TECTORIAL MEMBRANE
EYE
EYESPOT
PINHOLE CAMERA EYE
COMPOUND EYE
CAMERA LENS EYE
OMMATIDIUM
RETINA
ROD
CONE

The most primitive nervous system is the diffuse *nerve net* found in the Cnidaria. The neurons are scattered throughout the body. The cnidarian response to stimuli is general as there is no centralization of the system. The Platyhelminthes show a nervous system with *ganglia,* clusters of nerve cell bodies; that is, the beginning of specialization and centralization of nervous function. In the Annelida, the specialization and centralization increases. The ganglia in the head begin to take on the function of a brain. The nervous system becomes much more complex in the vertebrates. There is a clear CNS and a clear *peripheral nervous system* (*PNS*). The CNS is further specialized into a brain and a spinal cord. The brain and spinal cord are usually encased in either cartilage or bone. The brain and spinal cord are usually covered by three layers of membranes called the *meninges*. The PNS constitutes all of the nervous system that is not part of the CNS—the *nerves*, bundles of axons, and ganglia.

The CNS is subdivided into the brain and spinal cord. The brain can be further divided into the *forebrain, midbrain,* and *hindbrain.* The forebrain itself is divided into the *diencephalon,* which is composed of the *thalamus, hypothalamus,* and the *posterior lobe* of the *pituitary gland,* and the *telencephalon,* which is composed primarily of the *cerebrum.* During the evolution of the vertebrates, various segments and functions of the forebrain increase and shift. For example, in fishes, the thalamus is responsible for the relay of information from areas of olfactory stimulation to the midbrain, but in higher vertebrates, the thalamus becomes a center for integration of sensory information. The midbrain is primarily composed of the *optic lobe.* The hindbrain is composed of the *cerebellum,* the *pons,* and the *medulla oblongata.*

The PNS is subdivided into the *somatic nervous system,* that portion that controls the skeletal muscles of the body, and the *autonomic nervous system,* that portion that controls smooth muscles and glands. The somatic nervous system generally concerns itself with stimuli from the "outside" of the body while the autonomic nervous system generally concerns itself with stimuli from "within" the body, and is therefore involved with the maintenance of homeostasis. The autonomic nervous system is composed of two parts, the *sympathetic system,* which prepares the body during stress situations, and the *parasympathetic system,* which acts antagonistically to the sympathetic system.

In humans, there are 12 pairs of *cranial nerves* that convey information both from sense receptors to centers in the CNS and away from the CNS, to effectors such as glands or muscles. There are also 31 pair of *spinal nerves* with similar function to the cranial nerves.

Sense Receptors

Among the specializations in the nervous system observed during evolution, there are the developments of a number of sense receptors. Internal receptors sense various physical and chemical changes in blood and tissues while external receptors sense various physical and chemical changes in the external environment. We can classify receptors as *photoreceptors,* those that detect visible light stimuli; *chemoreceptors,* those that detect chemical stimuli; *mechanoreceptors,* those that detect mechanical energy stimuli; and *thermoreceptors,* those that detect thermal stimuli. Among the general sense receptors in higher animals are those in skin for the detection of pressure, pain, touch, heat, and cold, and those receptors in the joints and muscles, the *proprioceptors.* Gravity receptors in the invertebrates are called *statocysts,* while in the vertebrates they are called *hair cells.* The latter are pressure receptors in the lateral line organs of fishes, orientation to the pull of gravity detectors in the inner ears of vertebrates, and sound detectors in the inner ear of vertebrates.

The Ear

Among the special sense receptors in higher animals are the *eye* and the *ear.* Let us first consider the ear of higher vertebrates. The outer portion of the ear is composed of a *pinna,* the funnel; the *external auditory meatus,* the ear canal; and the *tympanic membrane,* the ear drum. The middle portion of the ear is composed of the *malleus, incus,* and *stapes,* three small bones, with the malleus pressed against the tympanic membrane and the stapes pressed against another membrane called the *oval window*. The pressure in this middle-ear chamber is stabilized through a tube, the *eustachian tube,* which opens into the back of the throat. The inner portion of the ear is composed of a complex structure that includes three *semicircular canals,* two saclike structures—the *saccule* and the *utricle*—and a snail-shell-like structure called the *cochlea*. The semicircular canals are organs of equilibrium. Each canal lies in a different plane from the other two and can detect acceleration of the head in a particular direction. The saccule and utricle contain calcium carbonate crystals, *otoliths,* which are embedded in a gelatinous material and stimulate hair cells. These two structures detect position with respect to gravity.

The cochlea is a coiled tubular structure that contains three canals separated by membranes. The *vestibular canal* and the *tympanic canal* both contain a fluid called *perilymph,* and they sandwich the third canal, the *cochlear duct*. The cochlear duct contains *endolymph*. It also contains a complex structure known as the *organ of Corti* which is sandwiched between a *basilar membrane* and a *tectorial membrane*. The organ of Corti consists of hair cells that detect sound energy.

Normal hearing is the result of the mechanical energy of sound being funneled into the external auditory meatus where it creates mechanical vibration of the tympanic membrane. The membrane in turn sets up mechanical vibrations in the malleus, incus, and the stapes. The stapes sets up mechanical vibrations in the oval window which are then translated to the perilymph in the vestibular canal. The vibrating perilymph establishes sympathetic vibrations in the endolymph of the cochlear duct and thereby stimulates the hair cells in the organ of Corti. At this point, the mechanical energy stimulates a wave of depolarization that moves to the auditory centers of the brain.

The Eye

The first photoreceptors were *eyespots,* detectors of differences in light intensity, that evolved in various forms in the protists, platyhelminthes, and annelids. The image-forming eye is capable of developing a true image. There are three basic types of image-forming eye in animals: the *pinhole camera* type, the *compound eye* type, and the *camera lens* type. The pinhole-type eye is found in many invertebrates and is limited to the amount of visual information that can enter the eye. The compound eye is a multifaceted structure found in some arthropods. It develops an image by allowing light from different angles to enter different photoreceptor units, *ommatidia*. The camera lens-type eye allows light to enter through an adjustable lens which focuses the incoming image on a receptor surface, the *retina*. Specialized receptor cells in the retina, the *rods* and *cones,* are sensitive to some property of light and synapse with dendrites of sensory neurons. The rods are sensitive to low light and enable the eye to see in these conditions. The cones are receptors that allow some animals to see color.

Miniquiz on the Nervous System

1. The compound eye of many insects consists of hundreds of visual units known as

   (A) nephridia
   (B) nephrons
   (C) ommatidia
   (D) mitochondria
   (E) cones

2. The three bones of the middle ear are the

   (A) malleus, incus, and patella
   (B) malleus, incus, and stapes
   (C) malleus, stapes, and scapula
   (D) stapes, clavicle, and patella
   (E) ilium, ischium, and pubis

3. The utricle and saccule are part of the

   (A) kidney
   (B) outer ear
   (C) middle ear
   (D) digestive system
   (E) inner ear

4. The neuron process that conducts impulses toward the cell body is the

   (A) dendrite
   (B) synapse
   (C) axon
   (D) nephron
   (E) none of these

5. Color vision in the human eye is possible because of the presence of the

   (A) rods
   (B) pupil
   (C) cones
   (D) sclera
   (E) ommatidia

6. Individual neurons communicate with other neurons through a nonprotoplasmic gap known as the

   (A) node
   (B) synapse
   (C) geopole
   (D) axon
   (E) synapsis

7. The centers for memory, learning, reasoning, and many other faculties such as sight, hearing, and touch are located in the

   (A) cerebellum
   (B) medulla oblongata
   (C) cerebrum
   (D) thalamus
   (E) spinal cord

8. Nerve impulses are the result of a wave of depolarization created by the movement of

   (A) hydrogen and aluminum ions
   (B) iron and magnesium ions
   (C) calcium and carboxyl ions
   (D) sodium and potassium ions
   (E) hydrogen and nitrogen ions

9. The elastic membrane at the end of the external auditory meatus is the

   (A) tympanic membrane
   (B) round window
   (C) oval window
   (D) semicircular canal
   (E) stapes

10. The special fatty-layered connective tissue that insulates an axon by wrapping itself around an axon is produced by a

    (A) Kupffer's cell
    (B) Henle cell
    (C) Schwann cell
    (D) Goblet cell
    (E) Alpha cell

11. After the depolarization of an axon membrane, the resting potential resulting from the asymmetrical distribution of ions across the membrane is reestablished by
    (A) osmosis
    (B) passive transport
    (C) diffusion
    (D) the sodium–potassium pump
    (E) dialysis

12. The most common neurotransmitter substance found at the synapse is
    (A) estrogen
    (B) acetylcholine
    (C) ptylin
    (D) pepsin
    (E) rennin

13. The fluid that is located in the cochlear duct is
    (A) perilymph
    (B) chyme
    (C) endolymph
    (D) aqueous humor
    (E) chyme

14. Destruction of the tectorial membrane is most likely to interfere with
    (A) digestion
    (B) vision
    (C) the sense of smell
    (D) taste
    (E) hearing

15. The most primitive nervous system is found in the members of the phylum
    (A) Cnidaria
    (B) Chondrichthyes
    (C) Annelida
    (D) Echinodermata
    (E) Chordata

16. The level of depolarization that must be exceeded in order to create a wave of depolarization is called the
    (A) resting potential
    (B) impulse
    (C) threshold
    (D) inhibition flux
    (E) enzyme deficiency

Answers and Explanations for the Miniquiz

1. **(C)** is correct. (A) are excretory structures found in the annelids. (B) are the functional units found in the kidney. (D) are the energy-producing organelles found in cells. (E) are color sensors in the retina.

2. **(B)** is correct. (A) The patella is the bone that forms the kneecap. (C) The scapula is the bone that forms the shoulderblade. (D) The clavicle is the bone commonly called the collarbone. (E) are the fused bones of the pelvis.

3. **(E)** is correct. These are parts of the inner ear along with the cochlea and the semicircular canals.

4. **(A)** is correct. (B) is the physical space between the end membrane of an axon and the membrane of a dendrite of another neuron, the cell body membrane of another neuron, or the membrane of a muscle cell. (C) conducts impulses away from the cell body of a neuron. (D) is the functional unit of the kidney.

5. **(C)** is correct. (A) detects light intensity differences. (B) is the space surrounded by the iris. (D) is the tough outer connective tissue layer of the eyeball. (E) are units of the compound eye.

6. **(B)** is correct. (A) is the region between the membranes of adjacent Schwann cells wrapped around an axon. These cells form the myelin sheath. (C) is nonsense. (D) is the cell process that conducts impulses away

from the cell body. (E) is the process of tetrad formation in metaphase I of meiosis.

7. **(C)** is correct. (A) is the brain structure that coordinates equilibrium and movement. (B) controls visceral reflex, and connects the brain and spinal cord. (D) is the area of sensory integration. (E) is an area of simple reflex.

8. **(D)** is correct. (A, B, C, and E) have nothing to do with impulses.

9. **(A)** is correct. (B, C, and D) are all parts of the inner ear. (E) is a bone of the middle ear.

10. **(C)** is correct. (A) is a cell type that is found in the liver. (B) is nonsense. (D) is a cell type found in the digestive tract. (E) is a cell type found in the pancreas.

11. **(D)** is correct. (A, B, C, and E) occur along concentration gradients. Reestablishment of the resting potential requires movement of molecules against concentration gradients and is therefore active transport.

12. **(B)** is correct. (A) is a steroid hormone produced in the ovary and the adrenal cortex. (C) is the carbohydrate-digesting hormone found in saliva. (D and E) are the protein-digesting enzymes produced in the stomach.

13. **(C)** is correct. (A) is the fluid found in the scala tympani and the scala vestibuli of the cochlea. (B) is the mixture of digesting food, saliva, and gastric juice that is developed in the stomach and moved into the duodenum. (D) is the alkaline watery fluid found in the anterior cavity of the eye. (E) is the partially digested mass that moves from the stomach to the duodenum.

14. **(E)** is correct. The tectorial membrane caps the organ of Corti in the cochlea, therefore its destruction would affect hearing.

15. **(A)** is correct. The Cnidaria have a primitive nerve net. There is no center for coordination.

16. **(C)** is correct. (A) is a condition that exists when the membrane is polarized. (B) is a wave of depolarization. (D and E) are nonsense.

THE ENDOCRINE SYSTEM

Words To Be Defined

HORMONE
PEPTIDE HORMONE
CYCLIC-AMP
STEROID HORMONE
ENDOCRINE GLAND
PITUITARY GLAND
INFUNDIBULUM
THYROID GLAND
THYROGLOBIN
THYROXIN
PARATHYROID GLAND
PARATHORMONE
ADRENAL CORTEX
ADRENAL MEDULLA
CORTISOL
TESTIS
OVARY
TESTOSTERONE
ESTROGEN
PROGESTERONE
PANCREAS
INSULIN
GLUCAGON
ISLETS OF LANGERHANS
PROSTAGLANDIN

The endocrine system is composed of glands that produce internal secretions, called *hormones,* chemical messengers. The two different types of hormones produced by the endocrine glands are chemically peptides and steroids. The current research into the mechanism of hormone action suggests each general type of hormone has its own receptor sites on cell membranes. A *peptide hormone* (an amino acid, a polypeptide, or a protein) appears to require receptor sites on the membranes of cells that are influenced by that hormone. When the hormone binds to the receptor site, it stimulates the formation of *cyclic-AMP* (cAMP). The cAMP is made from ATP within the cell. The cAMP activates cellular enzymes that facilitate the normal functions of the cell. A *steroid hormone* (complex rings of carbon, hydrogen, and nitrogen) appears to pass through the lipid layer of the cell membrane because it is a small, lipid-soluble molecule. Once inside the cell, steroid hormone combines with receptor molecules to form a steroid–receptor complex that moves into the nucleus where it binds with a specific segment of DNA and stimulates a particular gene for protein synthesis.

Endocrine glands are ductless glands that secrete their hormones directly into the blood. The organs of the body that respond to the hormones are called the target organs. The major endocrine glands of the body are the pituitary gland (anterior and posterior portions), the thyroid gland, the parathyroid gland, the adrenal gland (cortex and medulla), the pancreas, the testes, and the ovaries.

The *pituitary gland* is a small sphere that is suspended from the base of the hypothalamus by a stalk called the *infundibulum*. The pituitary gland has an anterior portion that produces at least six different hormones and a posterior portion that releases two hormones. The hypothalamus controls the production and release of hormones by the anterior pituitary. Hypothalamic-releasing hormones are produced in the hypothalamus and transported to the anterior pituitary through a portal circulation. The hypothalamus itself produces the hormones released by the posterior pituitary. These hormones move along nerves that begin in the hypothalamus and end in the posterior pituitary. The names and chief functions of the pituitary hormones are best summarized in a table form.

Anterior Pituitary Hormones	General Functions
GROWTH HORMONE (GH)	Stimulates body growth in children
LACTOGENIC HORMONE (LTH)	Causes the mammary glands to grow and produce milk during and after pregnancy
THYROID-STIMULATING HORMONE (TSH)	Stimulates the thyroid gland to produce and release thyroxin
ADRENOCORTICOTROPIC HORMONE (ACTH)	Stimulates the adrenal cortex to produce and secrete its hormones
FOLLICLE-STIMULATING HORMONE (FSH)	Stimulates the production of egg and sperm in their respective organs of production
LUTEINIZING HORMONE (LH)	Stimulates the sex hormone production in the respective organs of egg and sperm production

Posterior Pituitary Hormones	General Functions
ANTIDIURETIC HORMONE (ADH)	Stimulates water retention by the kidney
OXYTOCIN	Stimulates uterine contraction, particularly associated with the contractions of labor

The *thyroid gland* is a mass of tissue located on the trachea just below the larynx. The gland is composed of follicles that contain *thyroglobin,* a storage form of *thyroxin,* the hormone of the thyroid gland. Thyroxin increases metabolic rates in most cells of the body by causing an increase in the numbers of respiratory enzymes and the uptake of oxygen.

The *parathyroid gland* is represented by four separate masses embedded in the posterior surface of the thyroid gland. This gland produces a hormone known as *parathormone* (parathyroid hormone [PTH]) which controls calcium ions and phosphate ions in the blood. Parathyroid hormone stimulates the gut absorption of calcium ions, stimulates the activity of osteoclasts in bone reabsorption, and stimulates phosphate excretion by the kidney.

There are two adrenal glands, each of which is near a kidney. The outer portion of the kidney is the *adrenal cortex,* while the inner portion is known as the *adrenal medulla.* The adrenal cortex is under the control of ACTH. It produces and secretes three different groups of hormones: the glucocorticoids, the mineralocorticoids, and the sex hormones.

The glucocorticoids are steroid hormones that generally bring on the synthesis of glucose from nonglucose substances. These hormones are therefore helpful to the body during periods of stress, and are known to counteract inflammatory responses. One of the best-known examples of a glucocorticoid hormone is *cortisol.*

The mineralocorticoids are steroid hormones that regulate the levels of sodium and potassium ions in the blood. These hormones promote kidney absorption of sodium and kidney elimination of potassium. One of the best-known examples of a mineralocorticoid is aldosterone.

The sex hormones estrogen and testosterone are produced in very small quantities. The function of the male sex hormone in the female or the female sex hormone in the male is not known.

The *testes* and *ovaries* each produce their own hormones. The testes produce the steroid hormone *testosterone,* which controls the development of secondary sex characteristics in the male. The ovaries produce the steroid hormones *estrogen* and *progesterone,* which control the development of secondary sex characteristics in the female.

The *pancreas* is a compound organ that produces both digestive juices and protein hormones. The hormones *insulin* and *glucagon* are produced in the *islets of Langerhans.* Insulin stimulates the removal of glucose in the blood while glucagon stimulates the addition of glucose to the blood.

The *prostaglandins* (PG) are a newly discovered group of lipid compounds made from cell membrane fatty acids and produced in many cells. They appear to be involved in many aspects of reproduction and development and there seem to be many functions in nonreproductive cells also. The formation of PG is prevented by aspirin and this may account for its therapeutic effects.

MINIQUIZ ON THE ENDOCRINE SYSTEM

1. The tissue that is specifically affected by a particular hormone is usually designated as the

   (A) THS tissue
   (B) endocrine tissue
   (C) target tissue
   (D) medulla oblongata
   (E) none of these

2. One example of a gland that produces a steroid hormone is the

   (A) heart
   (B) esophagus
   (C) liver
   (D) ovary
   (E) spleen

3. Which function is *not* related to hormones of the adrenal cortex?

   (A) Regulation of the absorption of water balance in the kidney
   (B) Regulation of the synthesis of glucose from nonglucose substances
   (C) Counteraction of inflammatory responses
   (D) Regulation of calcium absorption from the blood serum
   (E) Regulation of sodium and potassium ions in the blood

4. The hormone that stimulates the testes and ovaries to produce their sex cells is

   (A) PG
   (B) insulin
   (C) LTH
   (D) thyroxine
   (E) ACTH

5. A drug that has a deleterious effect on the hypothalamus would have a direct effect on the production of

   (A) insulin
   (B) glucagon
   (C) PG
   (D) estrogen
   (E) oxytocin

6. Estrogen is to the adrenal cortex as estrogen is to the

   (A) thyroid gland
   (B) testis
   (C) pancreas
   (D) ovary
   (E) anterior pituitary gland

7. Which statement is *untrue?*

   (A) The posterior pituitary gland produces both oxytocin and ADH.
   (B) The anterior pituitary is under the influence of releaser factors produced in the hypothalamus.
   (C) The hormones FSH, LH, GH, and ACTH are all steroid hormones produced in the anterior pituitary gland.
   (D) Estrogen is the hormone that causes the development of secondary female sex characteristics.
   (E) Endocrine glands produce their hormones within their cells and secrete their products directly into the bloodstream.

Answers and Explanations for the Miniquiz

1. **(C)** is correct. (A) is nonsense. (B) is the tissue that produces steroid or protein hormones. (D) is the base of the brain and connects it to the spinal cord.

2. **(D)** is correct.

3. **(D)** is correct. Regulation of calcium absorption is a function of PTH.

4. **(C)** is correct. (A) are lipid compounds made from cell membrane fatty acids and behave as hormones that control a number of cell

activities. (B) is the pancreatic hormone that stimulates the cellular uptake of glucose from the bloodstream. (D) is the thyroid hormone that increases cellular respiration in mitochondria. (E) is the anterior pituitary hormone that stimulates the adrenal cortex into activity.

5. (**E**) is correct. Both oxytocin and ADH are produced by the hypothalamus and travel along nerves into the posterior pituitary gland where they are released into the bloodstream. (A and B) are produced in the pancreas. (C) is produced in the anterior pituitary gland. (D) is produced in the adrenal cortex and the ovary.

6. (**D**) is correct. Estrogen is produced in both of these glands. (A) produces thyroxin. (B) produces testosterone. (C) produces insulin and glucagon. (E) produces FSH, LH, ACTH, TSH, LTH, and GH.

7. (**A**) is incorrect. ADH and oxytocin are produced in the hypothalamus and move along nerve fibers into the posterior pituitary gland from which they are released into the bloodstream.

THE RESPIRATORY SYSTEM

Words To Be Defined

RESPIRATION
CELLULAR RESPIRATION
INTERNAL RESPIRATION
EXTERNAL RESPIRATION
BREATHING
GILL
TRACHEAL SYSTEM
LUNG
RESPIRATORY PIGMENT
HEMOGLOBIN
HEMOCYANIN
NASAL PASSAGE
EXTERNAL NARE
INTERNAL NARE
PHARYNX
LARYNX
GLOTTIS
EPIGLOTTIS
VOCAL CORD
TRACHEA
BRONCHUS
BRONCHIOLES
ALVEOLI
SURFACTANT
THORACIC CAVITY
DIAPHRAGM
INSPIRATION
EXPIRATION

All cells require molecular oxygen and generate carbon dioxide during the production of ATP for biological work. The exchange of gases in an organism is known as *respiration*. *Cellular respiration* is the production of ATP within mitochondria where molecular oxygen is used and carbon dioxide is produced. *Internal respiration* is the exchange of gases between blood and cells. *External respiration* is the exchange of gases between air and the blood. *Breathing* is the entrance and exit of gases from specialized structures of external respiration, the lungs.

There are four main animal structures of external respiration; (1) the surface of the body, (2) specialized outpocketings of a portion of the body surface (*gills*), (3) the simple inpocketing of a branched tubule system formed from a portion of the body surface (the *tracheal system*); and (4) the complex inpocketing of a portion of the body surface (the *lung*). All respiratory surfaces must be moist in order for gases to move in and out of cells and blood. "Breathing" is a term applied to those animals that have the respiratory structure known as the lung.

Most of the vertebrates use lungs for gas exchange. The amphibian lung is a rather simple partitioned sac. Many amphibians use both lungs and moist skin to exchange gases with the environment. The lung of reptiles is partitioned in a more complex manner than that of amphibians. The lungs of birds and mammals are a complex arrangement of continually branching tubes that terminate in highly vascularized sacs.

Water is a poor carrier of dissolved oxygen. Most animals use a *respiratory pigment* (large proteins that pick up and hold oxygen in areas of high oxygen concentration and release oxygen in areas of low oxygen concentration) to move oxygen from one place to another in the body. In higher animals the respiratory pigment is usually some form of the iron-containing protein *hemoglobin*. Some lower animals use a respiratory pigment that is a copper-containing protein, *hemocyanin*.

The human respiratory system comprises:

1. *Nasal Passages*—narrow canals with convoluted walls. They open externally by means of *external nares* and internally into the nasopharynx (a chamber just beyond the soft palate) by means of *internal nares*. The nasal passages are lined with hairs and mucus that filter out debris particles in the air. The passages also warm the air on its way to the lungs.
2. *Pharynx*—the crossover organ for air and food.
3. *Larynx*—a triangular box-shaped structure set in the floor of the pharynx. The *glottis* is a variable-sized opening that allows air moving through the pharynx to move into the larynx. A flap of soft tissue known as the *epiglottis* normally prevents food being moved from the oral cavity into the esophagus from entering the larynx. The larynx (sometimes referred to as the "voice box") contains the elastic ligaments known as the *vocal cords*.
4. *Trachea*—a tubular structure just below the larynx. The soft-walled tube is kept open by means of a series of c-shaped cartilaginous rings that give the trachea its characteristic "lumpy" appearance. Ciliated mucous membrane cells and mucus work together to trap debris and move it back out of the trachea.
5. *Bronchus*—The trachea divides into a right and left bronchus, each continuing to branch into smaller and smaller tubes known as *bronchioles*. The bronchi have a structure similar to that of the trachea. As the bronchioles become smaller and smaller, they lose their cartilaginous rings.
6. *Alveoli*—small air pockets or sacs at the terminal ends of the smallest bronchioles. The average human lung has about 700 million alveoli that provide it with an equivalent of about 100 times the surface area of the skin. Alveoli are prevented from collapsing due to surface tension by a lipoprotein (*surfactant*) that reduces surface tension.

The mechanism of breathing can be best understood by remembering that (1) a continuous column of air with a uniform pressure fills all of the sacs and tubes of the respiratory system, and (2) the lungs are contained in a closed chest cavity known as the *thoracic cavity*. The top and sides of the thoracic cavity are surrounded by ribs and intercostal muscles while the bottom is formed by a large dome-shaped muscle known as the *diaphragm*. The thoracic cavity is a closed system that has the following pressure–volume relationship: $P \times V = K$. This simply states that the intrathoracic pressure is inversely proportional to the intrathoracic volume.

Inspiration (breathing in) is initiated when the level of carbon dioxide and hydrogen ion concentration in the blood passing through the medulla oblongata reaches a threshold and stimulates impulses either to the diaphragm alone or to the diaphragm and intercostal muscles. These muscles then contract and effectively increase the intrathoracic volume and thereby reduce the intrathoracic pressure. The atmospheric pressure in the lungs then causes the lungs to inflate.

Expiration (breathing out) begins when the stretched alveoli stimulate special receptors in the alveolar walls. The receptors send inhibitory impulses from the inflated lung to the medulla oblongata causing the medulla to cease stimulation of the muscles of the diaphragm and ribs. The diaphragm and rib muscles relax, effectively decreasing the intrathoracic volume and thereby increasing the intrathoracic pressure. Both the increased intrathoracic pressure and the recoil of the stretched lung tissue cause the lung to deflate.

MINIQUIZ ON THE RESPIRATORY SYSTEM

1. A variety of aquatic animals derive their oxygen from water by using the

   (A) lungs
   (B) stomata
   (C) tracheal tubes
   (D) book lungs
   (E) none of these

2. The openings of the tracheal system are known as

   (A) spiricles
   (B) bronchioles
   (C) external nares
   (D) stomata
   (E) glomeruli

3. The initiation of an inspiration begins in the

   (A) heart
   (B) medulla oblongata
   (C) hypothalamus
   (D) nasal passages
   (E) alveoli

4. One disadvantage of deriving oxygen from air and not water is that

   (A) oxygen diffuses faster in water than it does in air
   (B) the respiratory surfaces are continually drying out and must be kept moist by body secretions
   (C) the nitrogen in air is very reactive with moistened tissue
   (D) oxygen pressure changes in air are very erratic
   (E) air contains 100 times less oxygen than does an equal volume of water

5. Which statement is true for tracheal systems?

   (A) Tracheal systems do not move gases in the tubules.

(B) Tracheal systems do not require the participation of a circulatory system in gas exchange.
(C) The gas exchange surfaces in the tracheal tubes must be dry for rapid gas exchange to occur in active insects.
(D) Insects do not produce carbon dioxide and therefore the tracheal system is only needed for oxygen acquisition.
(E) A decrease in temperature speeds up the exchange of gases in the tracheal system.

6. The organ that is characterized by a series of c-shaped rings of cartilage is the

(A) pharynx
(B) diaphragm
(C) alveolus
(D) trachea
(E) larynx

7. The initiation of expiration occurs when the

(A) level of dissolved oxygen in the blood increases to a level 10 times that of normal
(B) alveoli walls stretched by the inspiration stimulate special receptors in the alveolar walls
(C) diaphragm muscle begins to contract in response to stimuli from the medulla oblongata
(D) intrathoracic pressure becomes equal to the atmospheric pressure within the lungs
(E) epiglottis closes over the glottis

Answers and Explanations for the Miniquiz

1. **(E)** is correct. The skin and gills are the structures of gas exchange in aquatic animals. (A, C, and D) are gas-exchange structures found in land animals. (B) are openings in the leaf epidermis that provide for gas exchange.

2. **(A)** is correct. (B) are air tubes within the lungs. (C) are the external openings that lead into the nasal passages. (D) are openings in the leaf epidermis that provide for gas exchange. (E) are the capillary tufts found within the Bowman's capsule of the kidney.

3. **(B)** is correct. (A, C, D, and E) have nothing to do with the initiation or continuation of respiration.

4. **(B)** is correct. (A, C, D, and E) are all incorrect responses.

5. **(B)** is correct. (A) Gases must move in the tracheal tubes for effective gas exchange. (C) All gas exchange surfaces must be moist. (D) All living organisms produce carbon dioxide. (E) Gas exchange speeds up with increased temperatures.

6. **(D)** is correct. The bronchioles also contain c-shaped rings of cartilage.

7. **(B)** is correct.

THE REPRODUCTIVE SYSTEM

Words To Be Defined

MITOSIS
FRAGMENTATION
BUDDING
GAMETE
FERTILIZATION
ZYGOTE
PRIMARY SEX CHARACTERISTIC
SECONDARY SEX CHARACTERISTIC
LABIA MAJORA
LABIA MINORA
CLITORIS
HYMEN
FALLOPIAN TUBE
UTERUS
VAGINA
CERVIX
SCROTUM
TESTIS
PENIS
SEMINIFEROUS TUBULE
EPIDIDYMIS
VAS DEFERENS
SEMINAL FLUID
SEMINAL VESICLE
PROSTATE GLAND
COWPER'S GLAND
MENSTRUAL CYCLE
OVULATION
CORPUS LUTEUM
ESTROGEN
PROGESTERONE
ENDOMETRIUM
MENSTRUATION
PLACENTA
SPERMATOGENESIS
PRIMARY SPERMATOLYTE
HEAD
MIDPIECE
TAIL
OOGENESIS
PRIMARY OOCYTE OVUM
SECONDARY OOCYTE
FIRST POLAR BODY
SECONDARY POLAR BODY
CLEAVAGE
BLASTULA
YOLK
GASTRULA
ARCHENTERON
ECTODERM
MESODERM
ENDODERM
NEURULATION
NEURAL PLATE
ORGANOGENESIS
MORPHOGENESIS

Forms of Reproduction

The success of a species is determined by its ability to reproduce. Many "lower" animals reproduce by asexual means. Asexual reproduction is a conservative process because it does not need large numbers of eggs and sperm to assure the existence of the next generation. Asexual reproduction results in copies of the parent. The forms of asexual reproduction are *mitosis* (simple cell division in which diploid cells produce new diploid cells); *fragmentation* (small pieces of an organism break off and develop into a completely new animal); and *budding* (the animal forms a bud that develops into a new individual that then separates itself from the parent). Sexual reproduction is a wasteful process because it produces many more eggs and sperm than are needed for reproduction. Sexual reproduction is important because it involves genetic recombination and produces individuals that are more or less well adapted to environmental change than those produced by asexual reproduction. Sexually reproducing animals use male and female *gametes* (sex cells) in a process of *fertilization* (the process of egg-and-sperm fusion). After fertilization, the *zygote* (the first diploid cell of the next generation) divides numerous times, leading to embryonic development.

The human reproductive system is divisable into the internal sex organs and external or accessory sex organs. These organs are called *primary sex characteristics*. The *secondary sex characteristics* are the appearance and maintenance of pubic hair (males and females), underarm hair (males and females), and extensive facial and body hair (generally males).

The Human Female and Male

The primary external sex characteristics of the adult human female include the *labia majora* and *labia minora* (both are paired flaps of tissue that surround the openings to the vagina and urethra), the *clitoris* (the sensitive erectile tissue that is the analogue of the penis), and the *hymen* (a flap of tissue that partially covers the opening of the vagina). The internal sex organs include the ovaries (specialized organs in which meiosis results in the formation of egg cells), the *fallopian tubes* (ducts that convey eggs from the ovary to the uterus), the *uterus*, a large, hollow muscular organ in which the developing embryo embeds itself, and the *vagina*, the muscular tube that receives the penis for internal deposition of sperm. The external opening of the uterus (the *cervix*) is formed by a large sphincter muscle. The cervix protrudes into the uterus.

The primary external characteristics of the adult human male include the *scrotum* (double pouch of skin that surrounds two egg-shaped testes), the *testes* (specialized organs in which meiosis results in the formation of sperm cells), and the *penis* (a muscular organ with sinuses that can become engorged with blood leading to erection). The testis is composed of a coiled mass of *seminiferous tubules* in which spermatogenesis occurs. Above each testis is another coiled tubule (the *epididymis*) that stores mature sperm. Internally, each epididymus leads to a *vas deferens* (a tube that conducts sperm from the epididymis to the urethra). In addition, there are three glands that contribute to the formation of *seminal fluid* (an alkaline fluid that protects sperm cells against the hostile acid environment of the vagina). These glands are the *seminal vescicles*, the *prostate gland*, and the *Cowper's gland*.

Hormones play an important role in the *menstrual cycle* (the cyclic changes within the uterus that prepare the uterus for pregnancy and cause the uterine breakdown if there is no pregnancy). In humans, the typical menstrual cycle takes 28 days before the cycle repeats itself. The cycle begins when the anterior pituitary gland secretes increasing quantities of follicle stimulating hormone (FSH). The FSH causes the follicle to mature in preparation for egg release. A buildup of the anterior pituitary luteinizing hormone (LH) causes an egg to be released (*ovulation*) 14–16 days after the initiation of the cycle. The LH also causes the ruptured follicle to develop into the *corpus luteum*, a mass of tissue that secretes *estrogen* (the female hormone) and *progesterone* (the hormone of pregnancy). These hormones promote thickening of the *endometrium* (the vascularized lining of the uterus) and feed back to the anterior pituitary gland, inhibiting the production of FSH and LH. About 10 days after ovulation and its formation, the corpus luteum begins to degenerate, leading to a fall in progesterone. In response to this decrease in hormone the endometrium partially sloughs off (*menstruation*) and partially resorbs into the underlying tissue.

Successful fertilization of an egg will lead to embryonic development as the mass of tissue moves down the fallopian tube. About 10 days after embryonic development begins, the fetus implants in the endometrium of the uterus. In this case, the corpus luteum continues to produce progesterone and thereby prevents the destruction of the endometrium that is providing nourishment for the developing fetus. Some time around the fourth month of pregnancy the *placenta* (an important organ of pregnancy) is produced, partly by the endometrium and partly by the embryo.

Gamete Formation

Spermatogenesis (the process of sperm development) is essentially a two step process. First, the *primary spermatocyte* (the diploid cell destined to produce four sperm cells) goes through meiosis, the process that produces four haploid cells from a single diploid cell. Second, the haploid cells differentiate into a typical sperm cell with a *head* (nuclear-containing segment); a *midpiece* (mitochondrial-containing portion); and a *tail* (microtubule-containing flagellum). Spermatogenesis, once started, usually goes to completion. This is not the case with oogenesis.

Oogenesis (the process of egg development) is somewhat different from spermatogenesis. First, the *primary oocyte* (the diploid cell destined to produce a single haploid egg or *ovum*) goes through the first meiotic division to produce a *secondary oocyte* and a *first polar body* (a tiny diploid cell that is nonfunctional and eventually disintegrates). In some animals the first polar body may undergo a second division producing two nonfunctional haploid *secondary polar bodies* that eventually disintegrate. Once initiated, oogenesis usually does not go to completion but rather stops somewhere along the line. The stopping point varies in different animals. In the human, the secondary oocyte is the structure released from the follicle. The contact of a sperm cell with the oocyte stimulates the oocyte to divide once again, producing a haploid, nonfunctional, secondary polar body and a haploid ovum.

Fertilization and Development

The fusion of the haploid nuclei of the egg and sperm to form the zygote (the first diploid cell of the new generation) is called fertilization. In many aquatic animals fertilization is external, and in land animals and some aquatic animals it is internal. Following fertilization, the complex processes of mitosis, cell differentiation, and cell movement lead to the development of an organism that is characteristic to the species. There are generally four main stages of embryonic development. The first stage is *cleavage,* the production of a large number of cells by rapid cell division and the segregation of different cytoplasmic components into different groups of cells. The distribution of these components will influence the later specialization of the cells. The end result of cleavage is a hollow ball of cells called the *blastula*. The pattern of cleavage varies and often depends on the amount of stored food (*yolk*) in the original ovum.

The second stage in development is *gastrulation*, the rearrangement of the single-layered ball of cells into a three-layered mass of cells with a new central cavity known as the primitive gut or *archenteron*. The outermost layer is the *ectoderm* and is destined to give rise to the skin, brain, and spinal cord. The middle layer is the *mesoderm* and is destined to give rise to many of the internal organs, blood vessels, and muscles of the body. The innermost layer is the *endoderm* and is destined to give rise to the lining of the digestive tract.

The third stage is that of *neurulation*, the formation of the neural tube that is destined to become the brain and spinal cord. This process begins with the formation of the *neural plate*, a thickening of the ectoderm on a specific portion of the embryo. It establishes the symmetry in the animal.

The fourth stage in development is that of *organogenesis*, development of the organs of the body. There is strong evidence from experimentation with many kinds of embryos that the early pattern of embryonic development is set down in the cytoplasm of the ovum even before fertilization. After gastrulation, however, it appears that the full complement of zygote genes takes over the control of development. Some embryos are normally subject to *morphogenesis*, a change in body form during development. Morphogenesis is under the control of specific hormones.

MINIQUIZ ON REPRODUCTION

1. The blastocoele, the space in the center of the blastula, is replaced during gastrulation by the

(A) ectoderm
(B) archenteron
(C) blastospore
(D) zygote
(E) endoderm

2. Morphogenesis, a change in form in certain animals, is initiated by

(A) hormones
(B) ADH
(C) a dietary change
(D) increased heartbeat
(E) the contact of sperm with an egg

3. The structure that conducts the egg from the ovary to the uterus is the

(A) fallopian tube
(B) Cowper's gland
(C) vagina
(D) seminiferous tubule
(E) urethra

4. The correct path for sperm in the male reproductive tract is

(A) seminiferous tubule—urethra—epididymis—vas deferens
(B) seminiferous tubule—epididymus—vas deferens—urethra
(C) epididymus—vas deferens—urethra—seminiferous tubule
(D) vas deferens—seminiferous tubule—epididymus—urethra
(E) vas deferens—epididymus—seminiferous tubule—urethra

5. The structure(s) that does(do) not originate from the mesoderm is (are) the

(A) liver

(B) brain
(C) lungs
(D) muscles of the stomach
(E) skeletal muscles

6. The mesoderm is an embryonic layer that appears during

(A) cleavage
(B) fertilization
(C) gastrulation
(D) ovulation
(E) morphogenesis

7. Species that reproduce by means of internal fertilization are also characterized by

(A) the development of the embryo in water outside of the body
(B) a wide range of care patterns for the young after birth
(C) parental noninterest after the 17th week of development
(D) diminished potency with each new brood of young
(E) none of these

8. The periodic partial disintegration and discharge of the uterine lining is called

(A) pregnancy
(B) menstruation
(C) fertilization
(D) lactation
(E) gastrulation

9. The maturing human embryo is surrounded by an inner membrane known as the

(A) amnion
(B) yolk sac
(C) chorion
(D) notochord
(E) allantios

10. Most mammalian embryos are attached to the placenta by means of a/an

(A) umbilical cord
(B) endoderm
(C) clavichord
(D) chorion
(E) notochord

11. The process of cleavage terminates in the formation of a hollow ball of cells known as the

(A) blastula
(B) ovum
(C) gastrula
(D) archenteron
(E) zygote

12. During the process of spermatogenesis

(A) a diploid cell produces two new diploid cells that are capable of fertilizing a haploid egg
(B) a diploid cell produces four functional sperm cells that are capable of fertilizing a haploid egg
(C) four haploid cells fuse to form a tetrapoid zygote
(D) the primary spermocyte is produced by a normal meiosis
(E) none of these

Answers and Explanations for the Miniquiz

1. **(B)** is correct. (A) is the outer germ layer in the gastrula. (C) is the opening into the blastocoele, the hollow space in the blastula. (D) is the new diploid cell formed by the fusion of the egg and sperm. (E) is the innermost of the three primary germ layers.

2. **(A)** is correct. (B) is a posterior pituitary hormone. (C, D, and E) are nonsense.

3. **(A)** is correct. (B) is a gland that contributes to the formation of seminal fluid. (C) is the organ that receives the penis during internal fertilization. (D) is the structure where spermatogenesis occurs. (E) is the tube that serves the reproductive system in the male and the excretory system in both the male and the female.

4. **(B)** is correct.

5. **(B)** is correct. The ectoderm gives rise to the skin, brain, and spinal cord. (A, C, and E) are all derived from the mesoderm. (D) is derived from the endoderm.

6. **(C)** is correct. (A) is the process that changes the zygote into a multicellular hollow ball known as the blastula. (B) is the process of egg and sperm fusion. (D) is the process of egg release from the ovary. (E) is the process of form change.

7. **(E)** is correct.

8. **(B)** is correct. (A) is the condition of carrying a developing embryo within the body. (C) is the process of egg and sperm fusion. (D) is the process of milk production in mammals. (E) is the process which changes the single-layered blastula into a three-germ-layered gastrula.

9. **(A)** is correct. (B) is the membrane that surrounds the stored food for the embryo. (C) is the membrane that forms the outer membrane around the embryo and contributes to the formation of the placenta. (D) is the extraembryonic membrane that serves as a repository for nitrogen wastes produced by the embryo. (E) is the support rod that in developing vertebrates changes into the vertebral column.

10. **(A)** is correct. (B) is the inner primary germ layer of the gastrula that gives rise to the lining of the digestive system. (C) is nonsense. (D) is the membrane that forms the outer membrane around the embryo and contributes to the formation of the placenta. (E) is the support rod that in developing vertebrates changes into the vertebral column.

11. **(A)** is correct. (B) is the mature egg cell. (C) is the ball of cells composed of the three-germ layers. (D) is the primitive gut found in the gastrula. (E) is the new diploid cell of the next generation.

12. **(B)** is correct. (A, C, and D) are nonsense.

BEHAVIOR

Words To Be Defined

BEHAVIOR
INNATE BEHAVIOR
LEARNED BEHAVIOR
STEREOTYPED BEHAVIOR
REFLEX
SIGN STIMULUS
LEARNING
HABITUATION
CONDITIONED REFLEX
TRIAL AND ERROR
LATENT LEARNING
INSIGHT LEARNING
ENDOGENOUS
CIRCADIAN RHYTHM
ANNUAL RHYTHM
HORMONE
PHEROMONE
ALLOMONE
AWAKE
SLEEP
THIRST
SWEATING
MATING
CURIOSITY
AGGRESSION
SOCIAL BEHAVIOR
DOMINANCE HIERARCHY
TERRITORIALITY

INNATE VERSUS LEARNED BEHAVIOR

Behavior may be defined as an organism's outwardly directed activities in response to stimuli. *Innate* or *Instinctive behavior* is genetically determined and is difficult to alter because it is intrinsic. *Learned behavior* is environmentally determined and is acquired, altered, or eliminated as a result of experiences because it is extrinsic. In the early 1940s, a controversy arose among behaviorists; some claimed that behavior was the result of "nurture" while others claimed that behavior was the result of "nature." Today, most behaviorists believe that animals inherit a range of behavior patterns and that the environment determines what portion of the range will be expressed. Usually, when behavior patterns are produced nearly perfectly the first time that the organism is exposed to a particular stimulus, the behavior is innate. Learned behavior requires the expenditure of time, energy, and often repetition before the behavior is produced nearly perfectly upon stimulation. Many behavior patterns become programmed into the nervous system and become *stereotyped behavior* (acts that are always performed in near identical fashion). The simplest type of this kind of behavior is the *reflex* (an involuntary response to a stimulus). Stereotyped behavior requires a *sign stimulus* (also called a *releaser*), that portion of a mixed stimulus that elicits the behavior pattern.

Stimuli may initiate drive behavior, territorial behavior, conflict behavior, courtship behavior, or learning behavior. *Learning* is an enduring change in behavior in light of past experience. It is important in that it produces adaptive changes in some behavior patterns. Types of learning include: *habituation*, the simplest form of learning, in which the magnitude of the behavior response gradually decreases as the organism becomes accustomed to a particular stimulus; *conditioned reflex*, the response to a normally neutral stimulus (one that the animal would normally ignore as unimportant) that has become associated with some expected behavior; *trial and error*, a random response with either a reward or penalty associated with the behavior; *latent learning*, learning that occurs without any obvious behavior at the time that it occurs; and *insight learning*, the form of reasoning that requires the integration of past experiences in arriving at a novel behavior in the solution of a novel problem.

BIOLOGICAL CLOCKS

Rhythmic activities are common in animals and plants. In some cases, these activities are clearly controlled either by the nervous system or by the endocrine system. It is not clear what the control is for some rhythmic events. Rhythms are often *endogenous*, within the animal, and not due to the environment. *Circadian rhythms* are 24-hour, endogenous cycles that affect many physiological processes. Some animals have endogenous *annual rhythms*, cycles that repeat on a yearly basis. One may think of a biological clock as a pendulum that undergoes a fixed frequency of periodic oscillations. It is most likely that biological clocks in animals are under different controls from biological clocks in plants. Current research suggests that biological clock mechanisms in animals interact with their nervous systems and endocrine systems. In plants, biological clocks seem to interact with plant hormones.

PHEROMONES AND ALLOMONES

Hormones are chemicals that are produced by one group of cells, a tissue, or an organ and have a regulatory effect on another group of cells, a tissue, or an organ. *Pheromones* are substances secreted by one individual and are capable of stimulating a physiological or behavioral response in another individual of the same species. These chemical substances have been most studied in insects. Pheromones have been found to act as sex attractants, synchronizers of reproduction, accelerators of reproductive maturation, trail markers, and territorial markers.

Allomones are chemicals released from an individual of one species that have an effect on an individual of another species. Some allomones function as attractants while others function as repellents.

BEHAVIORAL STATES

Simple organisms rely almost exclusively on exogenous stimuli to elicit a particular behavioral state. With increasing organism complexity, behavioral states are more conditions that are brought about by a complex interrelationship between endogenous and exogenous stimuli. The higher the degree of complexity in animals, the more the endogenous stimuli participate in eliciting a particular behavioral state. Some examples of behavioral states:

Awake A higher animal condition in which the organism is active and responsive to stimuli

Sleep A higher animal condition in which the organism is quiescent, relaxed, and unresponsive to stimuli

Thirst A condition in a land animal in which the animal seeks water.

Sweating A condition in a homeothermic animal in which the body seeks to regulate temperature

Mating A condition in lower animals that appears to be determined by innate factors rigidly controlled by the nervous system, and in higher animals that appears to be controlled by the higher centers of a complex central nervous system in coordination with learning. One important part of mating behavior is courtship. As many animals maintain a minimum distance from one another, even with others of their species, it is necessary for some behavior that allows a member of the opposite sex to come close enough for mating

Curious A condition of higher animals that may well be a complex of responsive types of behavior under a single name

Aggressive A condition of higher animals that results in hostile behavior, the two main forms of which are *threatening* and *violent*

SOCIAL ORGANIZATION

Social behavior is involved in a situation in which groups of individuals live in close proximity and communicate or cooperate with other members of the groups. Various roles may be assigned to individuals, on the basis of age, sex, size, or physical strength. The phenomenon of *dominance hierarchy*, the organization of social animals in a natural situation in which the more dominant animals control the less dominant ones by threatening and aggressive actions, is important. It reduces the amount and intensity of fighting and thereby allows members of the species to conduct activities that are of more value to the organization.

Territoriality, a behavior which stakes out an animal's territory, is also a common form of social organization. It allows the strongest and most dominant members of a species to establish choice living space and access to resources for survival and reproduction. In this way, the best adaptations are passed along from generation to generation. Social organization in humans is the most complex known among the animals.

MINIQUIZ ON BEHAVIOR

1. The term "behavior" is best defined as
   (A) the temporary storage of data in the brain
   (B) the repetition of a stimulus with gradually decreasing intensity
   (C) the outwardly directed activities in response to stimuli
   (D) a successive series of trials with decreasing numbers of incorrect responses
   (E) an enduring change in behavior in light of past experience

2. Biorhythms that are timed to approximately 24-hour cycles are termed
   (A) annual rhythms
   (B) circadian rhythms
   (C) tidal rhythms
   (D) episodes
   (E) centennials

3. The situation in which a young bird that has never participated in a migration navigates, by means of an unlearned response, to a new nesting ground for the species, is an example of
   (A) taxis
   (B) instinct
   (C) reflex
   (D) territoriality
   (E) kinesis

4. The social organization in animals in which some animals are dominant and others are subservient is known as
   (A) dominance hierarchy
   (B) instinctive behavior
   (C) territoriality
   (D) sign stimulus
   (E) endogenous rhythms

5. The behavior in homeotherms that seeks to regulate body temperature is
   (A) sweating
   (B) curious
   (C) mating
   (D) aggressive
   (E) sleeping

6. A chemical produced by one group of cells, a tissue, or an organ that has a regulatory effect on another group of cells, a tissue, or an organ, is generally called a/an
   (A) hormone
   (B) repellent
   (C) allomone
   (D) curiosity
   (E) pheromone

Answers and Explanations for the Miniquiz

1. **(C)** is correct. (A) is short-term memory. (B) is meaningless. (D) is trial and error. (E) is learning.

2. **(B)** is correct. (A) are rhythms that repeat each year. (C) are rhythms that do not occur every 24 hours. (D) are events that have a prescribed beginning and a prescribed ending. (E) are rhythms that repeat each 100 years.

3. **(B)** is correct. (A) is a movement in response to a stimulus. (C) is merely a movement. (D) is a behavior that stakes out a territory. (E) is a movement.

4. **(A)** is correct. (B) is a genetically determined behavior. (C) is a behavior that stakes out a territory. (D) is that portion of a mixed stimulus that elicites a behavior pattern. (E) are internal rhythms.

5. **(A)** is correct. (B) is a complex response to stimuli. (C) is a behavior that will assure reproduction. (D) is a hostile behavior elicited by a stimulus. (E) is a quiescent, relaxed behavior.

6. **(A)** is correct. (B) is a type of allomone. (C) is a substance released by one organism that effects another of a different species. (D) is a form of behavior. (E) is a substance released by one organism that affects another of the same species.

ECOLOGY

Words To Be Defined

ENVIRONMENT
PHYSICAL ENVIRONMENT
BIOTIC ENVIRONMENT
ECOLOGY
NICHE
HABITAT
ADAPTATION
LONGEVITY ADAPTATION
REPRODUCTION ADAPTATION
MALADAPTATION
POPULATION
GROWTH RATE
CARRYING CAPACITY
MORTALITY RATE
AGE STRUCTURE
COMMUNITY
COMPETITION
PREDATION
SYMBIOSIS
PRINCIPLE OF COMPETITIVE EXCLUSION
NICHE OVERLAP
NICHE SHIFT
FOOD WEB
ECOSYSTEM
PARASITE
ECTOPARASITE
ENDOPARASITE
MUTUALISM
COMMENSALISM
PRODUCER
CONSUMER
PRIMARY CONSUMER
HERBIVORE
SECONDARY CONSUMER
TERTIARY CONSUMER
CARNIVORE
OMNIVORE
BIOMASS
DECOMPOSER
PRODUCTIVITY
NET PRODUCTIVITY
TROPHIC LEVEL
FOOD CHAIN
BIOGEOCHEMICAL CYCLE
WATER CYCLE
CARBON CYCLE
GREENHOUSE EFFECT
NITROGEN CYCLE
AMMONIFICATION
NITRIFICATION
ASSIMILATION
PHOSPHOROUS CYCLE
ECOLOGICAL SUCCESSION
CLIMAX COMMUNITY
BIOME
TUNDRA
TAIGA
TEMPERATE DECIDUOUS FOREST
TEMPERATE GRASSLANDS
SAVANNAS
CHAPARRAL
DESERT
TROPICAL RAIN FOREST

ADAPTATION TO THE ENVIRONMENT

That which is around an organism, the *environment*, is divisible into two components: the *physical environment* (nonliving matter and energy) and the *biotic environment* (living matter or organisms). The study of the interaction between organisms and their environments is called *ecology*. *How* an organism lives is known as its *niche*, while *where* an organism lives is known as its *habitat*. An *adaptation* is any adjustment to the environment that increases an organism's chances of survival (*longevity adaptation*) and reproduction (*reproductive adaptation*), while a *maladaptation* is the opposite.

The fundamental unit for ecological study is the *population*, a group of organisms sharing a common gene pool and a common locality. Four important characteristics of a population are *growth rate,* birth rate minus death rate plus or minus emigration and immigration effects; *carrying capacity,* the number of individuals of a species that can be supported by local resources; *mortality patterns,* patterns of death during stages of the life cycle common for that species; and *age structure,* the proportions of different individuals of different ages in the populations.

ENVIRONMENTAL INTERRELATIONSHIPS

A *community* comprises all interacting populations inhabiting a common environment. The interactions within a community are the major factors in natural selection and the determination of numbers of individuals in each population and numbers of species in the community. Three principal categories of community interaction are *competition,* the interaction between organisms using a common resource present in limited supply; *predation,* the interaction between organisms where one species is used as a resource by another species; and *symbiosis,* the interaction between two species usually in close and long-term association.

Generally, the more or less similar that two populations are, the more or less they will compete relatively speaking. Gause's *Principle of Competitive Exclusion* in essence states that if two species are in competition for a limited resource, one of the two species will eventually be eliminated. This principle predicts that only one species at a time can occupy a particular ecological niche. Where there are two similar species in apparent coexistence, it has been found that each actually occupies a slightly different niche from the other, and what appears to be a common resource is actually subdivided in such a way as to make coexistence possible. More than one species utilizing a common, limited resource is a situation of *niche overlap*. Natural selection may then contribute to an increase in the differences between the two species and thereby result in a *niche shift.* A niche shift minimizes the competition between the two species.

A *food web* is a complex set of interactions among organisms through which energy and matter move within a community or *ecosystem* (a community and the abiotic factors with which it interacts). A food web represents feeding relationships among various organisms in a community. The key relationship is that of predator and prey. Predator population size is often, but not exclusively, limited by prey availability. Prey population is also influenced by prey food supply. Predator–prey interactions may result in cycles of population numbers, increases in species diversity, and evolutionary change for the species involved.

In symbiosis, there are three possible relationships between two different populations. The first is *parasitism,* a relationship in which one species receives benefit at the expense of and harm to the other species. In certain respects, parasitism can be considered a special form of predation in which the predator is significantly smaller then the prey. Many parasites may harm the host in such a way as to cause the death of the host. This is generally not to the advantage of the parasite. *Ectoparasites* are parasites that attach themselves to the outside of the host, while *endoparasites* are those that invade the body of the host.

Endoparasites often invade those parts of the body that receive a rich blood supply. The second general symbiotic relationship between two different populations is *mutualism,* a relationship in which both species receive benefit. The third is *commensalism,* a relationship in which one species receives benefit but the other receives neither benefit nor harm.

ECOSYSTEMS

As stated earlier, an ecosystem consists of all of the biotic and abiotic components within a given area. Interactions in an ecosystem result in two important phenomena. First, there is a flow of energy from *producers* (autotrophic organisms that capture sunlight energy and convert it into chemical energy), to *consumers* heterotrophic organisms that derive energy from the consumption of autotrophs, or from the consumption of other heterotrophs. *Primary consumers* are *herbivores,* animals that eat plants. *Secondary* and *tertiary consumers* are *carnivores,* animals that eat animals. *Omnivores* are a combination of the two supplementing their diets with both plants and animals. Second, there is a cycling of material from the abiotic environment through the biotic environment (the bodies of plants and animals) and back again to the abiotic environment. Material in the biotic environment is called *biomass* (the total dry weight of all the organisms being measured). *Decomposers* are the organisms that assure the return of biotic material to the abiotic environment.

Productivity is a measure of the total amount of light energy of the sun converted to chemical energy in plants. *Net productivity* is the total amount of energy entering autotrophs minus the energy needed to operate metabolic activities, minus energy lost to entropy. It has been estimated that only 1–3% of the energy that falls on a plant is stored in organic molecules. As the energy moves from producer to primary consumer to secondary consumer, etc., there is a continual reduction in the energy available for storage. This reduction in energy is predicted by the *Second Law of Thermodynamics* which states that in all energy conversions the potential energy of the final state will always be less than the potential energy of the initial state. As energy passes from one organism to another it is considered to pass from one *trophic level* to the next trophic level. The first trophic level is composed of producers, the second is composed of primary consumers, the third is composed of secondary consumers, etc. The typical *food chain* (the sequence of organisms related to one another as predator and prey) is composed of three or four trophic levels. Although rare, there are food chains with five or six trophic levels. Biologists find it useful to describe levels of energy, biomass, and numbers of individuals, in terms of pyramids.

BIOGEOCHEMICAL CYCLES

Contrary to the one-way flow of energy through the ecosystem, substances such as carbon, nitrogen, phosporus, water, and oxygen cycle through the ecosystem.

These cycles are referred to as *biogeochemical cycles*. Four important cycles are the water cycle, the carbon cycle, the nitrogen cycle, and the phosphorus cycle.

The water cycle circulates water from the surface of the earth through evaporation into the atmosphere where it accumulates, comes down as rain, and enters living organisms. Water leaves organisms in a variety of ways and returns to the abiotic environment where it continues the cycle. Water is therefore continuously available for organisms.

The carbon cycle moves carbon from the atmosphere into the biotic environment through carbon dioxide fixation in the Benson–Calvin cycle. Once in the biotic environment, carbon moves from photosynthesis into organic structure. From the biotic environment, carbon moves back to the abiotic environment by either cellular respiration or combustion of organic materials. The combustion of fossil fuels such as coal and oil can increase carbon dioxide in the atmosphere. Increased carbon dioxide can produce a *greenhouse effect*, allowing solar heat to enter the earth's atmosphere but preventing the heat from reradiating into space, and thereby causing a heating up of the surface of the earth.

The nitrogen cycle is important because nitrogen is a necessary element in proteins and nucleic acids. Most organisms cannot take nitrogen out of the atmosphere. They rely on the limited supply of nitrogen in the soil. The three principal steps in the cycle are *ammonification*, *nitrification*, and *assimilation*. Ammonification is the process by which bacteria and fungi in the soil decompose organic materials and release excess nitrogen in the form of ammonia. Nitrification is the two-step process that changes ammonia into nitrates which can be absorbed and utilized by plants. The first step is the oxidation of ammonia into nitrites by members of one genus of bacteria. Next the nitrites are oxidized into nitrates by another genus of bacteria. Assimilation is the conversion of nitrate back into ammonia once it has entered the cell. The ammonia is then coupled to organic molecules to make amino acids, nucleosides, etc.

The phosphorus cycle is important, as phosphorus is an essential component of lipids, nucleic acids, and ATP. Phosphorus is constantly being made available to organisms by the erosion of rock. The major source, however, is concentrated in the tissues of organisms, and, therefore, the decaying bodies of living things are the primary source. Much of the phosphorus in the cycle in recycled by the decomposers.

ECOLOGICAL SUCCESSION

A sequence of organismic changes on a particular site is called *ecological succession*. These changes are the result of competitive interactions among the species on that site. Ecological succession is a regular, predictable process that has certain common features no matter where and how succession is proceeding. First, there is an increase in the total biomass of the site. Second, there is a decrease in the net productivity in relation to biomass. Third, mature systems have enhanced capacity to accumulate and hold nutrients within the system. Finally, there is an increase in the number of species. The terminating, relatively stable community in an ecological succession is called the *climax community*.

BIOMES

A very large climax community in a geographic region of one kind of climate is called a *biome*. There are eight major land biomes on the planet earth:

Tundra A form of grassland characterized by *permafrost*, a layer of permanently frozen subsoil

Taiga A conifer forest characterized by long, severe winters and a permanent cover of snow

Temperate Deciduous Forest A region with a warm growing season and moderate precipitation, alternating with cold periods poorly suited for growth

Temperate Grasslands A region of rolling to flat terrain, periodic droughts, and hot and cold seasons. It serves as a transition between a temperate forest and a desert, and is usually located within the interior of a continent

Savannas A region of tropical grassland with alternating wet and dry seasons. The grassland is dotted with clumps of trees

Chaparral A region dominated by small trees and spiny shrubs and subject to mild, rainy winters alternating with hot, dry summers

Desert A region of little rainfall, resulting from wind patterns mainly of falling warm air. A region is considered a desert if there is an annual rainfall of less than 25 centimeters

Tropical Rain Forest A region of an annual rainfall of between 200 and 400 centimeters

MINIQUIZ ON ECOLOGY

1. Interrelated groups of food chains are known as a

   (A) niche
   (B) food web
   (C) population
   (D) habitat
   (E) none of these

2. One important component of the carbon cycle is

   (A) molecular oxygen
   (B) molecular ATP
   (C) ammonia
   (D) water
   (E) molecular carbon dioxide

3. Animals that eat animals that eat plants are called

   (A) primary consumers
   (B) secondary consumers
   (C) producers
   (D) decomposers
   (E) tertiary consumers

4. How an organism lives in relationship to the other members of its biotic community or ecosystem is referred to as a/an

   (A) niche
   (B) habitat
   (C) food web
   (D) ecological succession
   (E) biome

5. Organisms that live on dead and decaying material are called

   (A) parasites
   (B) producers
   (C) primary consumers
   (D) decomposers
   (E) omnivores

6. A major share of nitrogen fixation in the soil is carried out by

   (A) producers
   (B) primary consumers
   (C) decomposers
   (D) bacteria
   (E) viruses

7. The basic pattern of an interaction between two species in which one benefits while the other neither benefits nor is harmed is called

   (A) commensalism
   (B) parasitism
   (C) mutualism
   (D) competition
   (E) neutralism

8. Bacteria play an important role in the

   (A) water cycle
   (B) carbon cycle
   (C) nitrogen cycle
   (D) phosphorous cycle
   (E) aluminum cycle

9. A region of grasslands dotted with clumps of trees, and alternating with wet and dry seasons is known as

   (A) tundra
   (B) taiga
   (C) savanna
   (D) desert
   (E) chaparral

10. Temperate grasslands are characterized as regions of

    (A) grassland with a permafrost layer
    (B) conifer forest which experience long, severe winters
    (C) rolling to flat terrain with periodic droughts and alternating hot and cold seasons
    (D) an annual rainfall of less than 25 centimeters as a result of wind patterns mainly of falling warm air
    (E) an annual rainfall of between 200 and 400 centimeters

11. Three principal categories of community interaction are

    (A) competition, predation, and parasitism
    (B) competition, symbiosis, and parasitism
    (C) predation, symbiosis, and mutualism
    (D) symbiosis, predation, and competition
    (E) mutualism, commensalism, and parasitism

12. A biome is a/an

    (A) heterotrophic organism that derives its energy from the consumption of autotrophs
    (B) group of interacting organisms sharing a single habitat
    (C) a small climax community in a region of one kind of temperature
    (D) a very large climax community in a geographic region of one kind of climate
    (E) sequence of organismic changes on a particular site

Answers and Explanations for the Miniquiz

1. **(B)** is correct. (A) characterizes how an organism lives. (C) is a group of organisms sharing a common gene pool and a common locality. (D) is where an organism lives.

2. **(E)** becomes fixed in the Benson–Calvin cycle and serves as the foundation for the carbon skeleton in organic molecules. (A) is part of the oxygen cycle. (B) is part of the phosphorus cycle. (C) is part of the nitrogen cycle. (D) is the molecule of the water cycle.

3. **(B)** is correct. (A) are animals that eat plants. (C) are photosynthetic organisms which produce organic molecules from inoganic molecules and sunlight energy. (D) are organisms that live on dead and decaying material. (E) are animals that eat animals that in turn eat animals that eat plants.

4. **(A)** characterizes how an organism lives in relation to other organisms. (B) characterizes where an organism lives. (C) describes the complex feeding relationships among many species interacting. (D) is a sequence of change in organisms on a particular site. (E) is a very large climax community in a geographic region of one kind of climate.

5. **(D)** is correct. (A) are organisms engaged in a form of symbiosis. (B) are photosynthetic organisms. (C) are herbivores. (E) are consumers that eat both plant and animals.

6. **(D)** fix nitrogen into nitrites and nitrites into nitrates. (A) are photosynthetic organisms. (B) are herbivores. (C) are organisms that live on dead and decaying organic matter. (E) are not cells and have no known beneficial function in nature.

7. **(A)** is correct. (B) is a relationship in which one organism benefits and the other organism is harmed. (C) is a relationship in which both participants benefit. (D) is the interaction between two organisms using the same resource. (E) is not a term used in ecology.

8. **(C)** is correct.

9. **(C)** is correct. (A) is a grassland that is characterized by permafrost. (B) is a conifer forest that is characterized by long, severe winters and a permanent snow cover. (D) is a region with an annual rainfall of less than 25 centimeters. (E) is dominated by small trees and spiny shrubs, and subject to mild, rainy winters alternating with hot dry summers.

10. **(C)** is correct. (A) describes a tundra. (B) describes a taiga. (D) describes a desert. (E) describes a tropical rain forest.

11. **(D)** is correct. Parasitism, mutualism, and commensalism are forms of symbiosis.

12. **(D)** is correct. (A) describes primary consumers. (B) is a competition. (C) is nonsense. (E) is an ecological succession.

PART FOUR

FINAL MODEL EXAMINATIONS

The examination that follows offers a second chance for you to assess your readiness for the Advanced Placement (AP) Examination in Biology. Allow yourself 90 minutes to answer all of the 120 multiple-choice questions in Section 1 and 90 minutes to answer the four essays in questions in Section 2. When you have completed the exam, check your answers against the Answer Key and Explanatory Answers at the end of the exam. Complete the Diagnostic Chart on page 213 to identify any remaining weaknesses.

Use the specially constructed answer sheet to record your answers for Section 1. Use plain or lined paper to answer the free-response questions in Section 2. Sample answers are provided for each essay question.

MODEL EXAMINATION 2, SECTION 1
ANSWER SHEET

1 Ⓐ Ⓑ Ⓒ Ⓓ Ⓔ	25 Ⓐ Ⓑ Ⓒ Ⓓ Ⓔ	49 Ⓐ Ⓑ Ⓒ Ⓓ Ⓔ	73 Ⓐ Ⓑ Ⓒ Ⓓ Ⓔ	97 Ⓐ Ⓑ Ⓒ Ⓓ Ⓔ
2 Ⓐ Ⓑ Ⓒ Ⓓ Ⓔ	26 Ⓐ Ⓑ Ⓒ Ⓓ Ⓔ	50 Ⓐ Ⓑ Ⓒ Ⓓ Ⓔ	74 Ⓐ Ⓑ Ⓒ Ⓓ Ⓔ	98 Ⓐ Ⓑ Ⓒ Ⓓ Ⓔ
3 Ⓐ Ⓑ Ⓒ Ⓓ Ⓔ	27 Ⓐ Ⓑ Ⓒ Ⓓ Ⓔ	51 Ⓐ Ⓑ Ⓒ Ⓓ Ⓔ	75 Ⓐ Ⓑ Ⓒ Ⓓ Ⓔ	99 Ⓐ Ⓑ Ⓒ Ⓓ Ⓔ
4 Ⓐ Ⓑ Ⓒ Ⓓ Ⓔ	28 Ⓐ Ⓑ Ⓒ Ⓓ Ⓔ	52 Ⓐ Ⓑ Ⓒ Ⓓ Ⓔ	76 Ⓐ Ⓑ Ⓒ Ⓓ Ⓔ	100 Ⓐ Ⓑ Ⓒ Ⓓ Ⓔ
5 Ⓐ Ⓑ Ⓒ Ⓓ Ⓔ	29 Ⓐ Ⓑ Ⓒ Ⓓ Ⓔ	53 Ⓐ Ⓑ Ⓒ Ⓓ Ⓔ	77 Ⓐ Ⓑ Ⓒ Ⓓ Ⓔ	101 Ⓐ Ⓑ Ⓒ Ⓓ Ⓔ
6 Ⓐ Ⓑ Ⓒ Ⓓ Ⓔ	30 Ⓐ Ⓑ Ⓒ Ⓓ Ⓔ	54 Ⓐ Ⓑ Ⓒ Ⓓ Ⓔ	78 Ⓐ Ⓑ Ⓒ Ⓓ Ⓔ	102 Ⓐ Ⓑ Ⓒ Ⓓ Ⓔ
7 Ⓐ Ⓑ Ⓒ Ⓓ Ⓔ	31 Ⓐ Ⓑ Ⓒ Ⓓ Ⓔ	55 Ⓐ Ⓑ Ⓒ Ⓓ Ⓔ	79 Ⓐ Ⓑ Ⓒ Ⓓ Ⓔ	103 Ⓐ Ⓑ Ⓒ Ⓓ Ⓔ
8 Ⓐ Ⓑ Ⓒ Ⓓ Ⓔ	32 Ⓐ Ⓑ Ⓒ Ⓓ Ⓔ	56 Ⓐ Ⓑ Ⓒ Ⓓ Ⓔ	80 Ⓐ Ⓑ Ⓒ Ⓓ Ⓔ	104 Ⓐ Ⓑ Ⓒ Ⓓ Ⓔ
9 Ⓐ Ⓑ Ⓒ Ⓓ Ⓔ	33 Ⓐ Ⓑ Ⓒ Ⓓ Ⓔ	57 Ⓐ Ⓑ Ⓒ Ⓓ Ⓔ	81 Ⓐ Ⓑ Ⓒ Ⓓ Ⓔ	105 Ⓐ Ⓑ Ⓒ Ⓓ Ⓔ
10 Ⓐ Ⓑ Ⓒ Ⓓ Ⓔ	34 Ⓐ Ⓑ Ⓒ Ⓓ Ⓔ	58 Ⓐ Ⓑ Ⓒ Ⓓ Ⓔ	82 Ⓐ Ⓑ Ⓒ Ⓓ Ⓔ	106 Ⓐ Ⓑ Ⓒ Ⓓ Ⓔ
11 Ⓐ Ⓑ Ⓒ Ⓓ Ⓔ	35 Ⓐ Ⓑ Ⓒ Ⓓ Ⓔ	59 Ⓐ Ⓑ Ⓒ Ⓓ Ⓔ	83 Ⓐ Ⓑ Ⓒ Ⓓ Ⓔ	107 Ⓐ Ⓑ Ⓒ Ⓓ Ⓔ
12 Ⓐ Ⓑ Ⓒ Ⓓ Ⓔ	36 Ⓐ Ⓑ Ⓒ Ⓓ Ⓔ	60 Ⓐ Ⓑ Ⓒ Ⓓ Ⓔ	84 Ⓐ Ⓑ Ⓒ Ⓓ Ⓔ	108 Ⓐ Ⓑ Ⓒ Ⓓ Ⓔ
13 Ⓐ Ⓑ Ⓒ Ⓓ Ⓔ	37 Ⓐ Ⓑ Ⓒ Ⓓ Ⓔ	61 Ⓐ Ⓑ Ⓒ Ⓓ Ⓔ	85 Ⓐ Ⓑ Ⓒ Ⓓ Ⓔ	109 Ⓐ Ⓑ Ⓒ Ⓓ Ⓔ
14 Ⓐ Ⓑ Ⓒ Ⓓ Ⓔ	38 Ⓐ Ⓑ Ⓒ Ⓓ Ⓔ	62 Ⓐ Ⓑ Ⓒ Ⓓ Ⓔ	86 Ⓐ Ⓑ Ⓒ Ⓓ Ⓔ	110 Ⓐ Ⓑ Ⓒ Ⓓ Ⓔ
15 Ⓐ Ⓑ Ⓒ Ⓓ Ⓔ	39 Ⓐ Ⓑ Ⓒ Ⓓ Ⓔ	63 Ⓐ Ⓑ Ⓒ Ⓓ Ⓔ	87 Ⓐ Ⓑ Ⓒ Ⓓ Ⓔ	111 Ⓐ Ⓑ Ⓒ Ⓓ Ⓔ
16 Ⓐ Ⓑ Ⓒ Ⓓ Ⓔ	40 Ⓐ Ⓑ Ⓒ Ⓓ Ⓔ	64 Ⓐ Ⓑ Ⓒ Ⓓ Ⓔ	88 Ⓐ Ⓑ Ⓒ Ⓓ Ⓔ	112 Ⓐ Ⓑ Ⓒ Ⓓ Ⓔ
17 Ⓐ Ⓑ Ⓒ Ⓓ Ⓔ	41 Ⓐ Ⓑ Ⓒ Ⓓ Ⓔ	65 Ⓐ Ⓑ Ⓒ Ⓓ Ⓔ	89 Ⓐ Ⓑ Ⓒ Ⓓ Ⓔ	113 Ⓐ Ⓑ Ⓒ Ⓓ Ⓔ
18 Ⓐ Ⓑ Ⓒ Ⓓ Ⓔ	42 Ⓐ Ⓑ Ⓒ Ⓓ Ⓔ	66 Ⓐ Ⓑ Ⓒ Ⓓ Ⓔ	90 Ⓐ Ⓑ Ⓒ Ⓓ Ⓔ	114 Ⓐ Ⓑ Ⓒ Ⓓ Ⓔ
19 Ⓐ Ⓑ Ⓒ Ⓓ Ⓔ	43 Ⓐ Ⓑ Ⓒ Ⓓ Ⓔ	67 Ⓐ Ⓑ Ⓒ Ⓓ Ⓔ	91 Ⓐ Ⓑ Ⓒ Ⓓ Ⓔ	115 Ⓐ Ⓑ Ⓒ Ⓓ Ⓔ
20 Ⓐ Ⓑ Ⓒ Ⓓ Ⓔ	44 Ⓐ Ⓑ Ⓒ Ⓓ Ⓔ	68 Ⓐ Ⓑ Ⓒ Ⓓ Ⓔ	92 Ⓐ Ⓑ Ⓒ Ⓓ Ⓔ	116 Ⓐ Ⓑ Ⓒ Ⓓ Ⓔ
21 Ⓐ Ⓑ Ⓒ Ⓓ Ⓔ	45 Ⓐ Ⓑ Ⓒ Ⓓ Ⓔ	69 Ⓐ Ⓑ Ⓒ Ⓓ Ⓔ	93 Ⓐ Ⓑ Ⓒ Ⓓ Ⓔ	117 Ⓐ Ⓑ Ⓒ Ⓓ Ⓔ
22 Ⓐ Ⓑ Ⓒ Ⓓ Ⓔ	46 Ⓐ Ⓑ Ⓒ Ⓓ Ⓔ	70 Ⓐ Ⓑ Ⓒ Ⓓ Ⓔ	94 Ⓐ Ⓑ Ⓒ Ⓓ Ⓔ	118 Ⓐ Ⓑ Ⓒ Ⓓ Ⓔ
23 Ⓐ Ⓑ Ⓒ Ⓓ Ⓔ	47 Ⓐ Ⓑ Ⓒ Ⓓ Ⓔ	71 Ⓐ Ⓑ Ⓒ Ⓓ Ⓔ	95 Ⓐ Ⓑ Ⓒ Ⓓ Ⓔ	119 Ⓐ Ⓑ Ⓒ Ⓓ Ⓔ
24 Ⓐ Ⓑ Ⓒ Ⓓ Ⓔ	48 Ⓐ Ⓑ Ⓒ Ⓓ Ⓔ	72 Ⓐ Ⓑ Ⓒ Ⓓ Ⓔ	96 Ⓐ Ⓑ Ⓒ Ⓓ Ⓔ	120 Ⓐ Ⓑ Ⓒ Ⓓ Ⓔ

MODEL EXAMINATION 2

SECTION 1

Time—90 minutes
Number of Questions—120
Percent of Total Grade—60

Directions: **For each of the 120 questions or incomplete statements below, select the choice from (A–E) that best answers the question or completes the statement. Record your answers on the answer sheet provided.**

1. Two examples of a monosaccharide are
   (A) sucrose and maltose
   (B) ribose and glucose
   (C) starch and glucogen
   (D) galactose and lactose
   (E) lactose and maltose

2. The structure that is not a fruit is the
   (A) peach
   (B) cucumber
   (C) apple
   (D) pear
   (E) none of these

3. Nine different blood proteins that become enzymatic when activated and which facilitate the breakdown of cell membranes are called
   (A) complement
   (B) antigens
   (C) plasma proteins
   (D) vaccines
   (E) antibody titer

4. Cloning provides support for the concept that
   (A) most cells contain the full range of genetic information for the individual even though each specialized cell expresses only a particular portion of that information
   (B) evolution by natural selection is based upon sexual reproduction
   (C) DNA is the genetic material while RNA is a complementary molecule to DNA
   (D) the cell is the basic unit of life, genetics, structure, function, heredity, and pathology
   (E) fraternal twins that arise from two different eggs are fertilized by two different sperm

5. The significance of mitosis can be summarized as
   (A) the cell division in which diploid cells produce haploid cells
   (B) a type of sexual reproduction in which diploid gametes are produced
   (C) the division in which diploid cells produce diploid cells that are genetic copies of the parent cell
   (D) a type of division during which synapsis occurs
   (E) a division which primarily occurs in the gonadal tissues

Question 6 is based on the following graph.

QUANTITY OF TISSUES IN THE HUMAN BODY

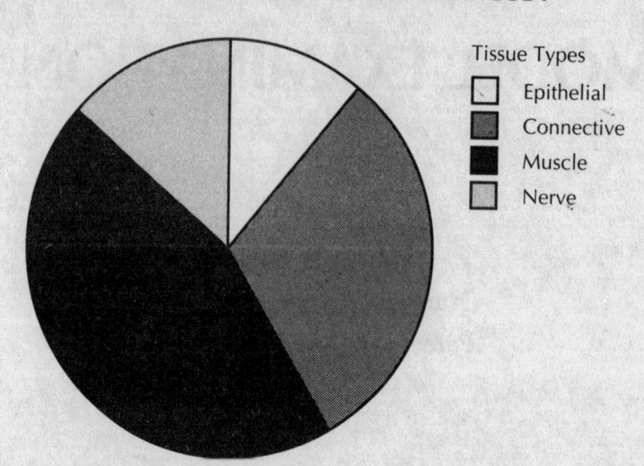

6. Based on the above graph, which of the following statements about tissue types is correct?
   (A) There is about 3 times as much nerve tissue present as epithelial tissue.
   (B) There is about 3 times as much connective tissue present as epithelial tissue.
   (C) There is about 5 times as much muscle tissue present as connective tissue.
   (D) There is about 3 times as much muscle tissue present as connective tissue.
   (E) both (A and B)

7. The triploid tissue in the developing seed is the
   (A) epidermis of the seed coat
   (B) endosperm
   (C) radicle
   (D) plumule
   (E) megaspore

8. One difference between man-made ecosystems and natural ecosystems is
   (A) man-made ecosystems are unstable
   (B) natural ecosystems are unstable
   (C) natural ecosystems produce more waste than do man-made ecosystems
   (D) man-made ecosystems are better able to handle wastes
   (E) man-made ecosystems have a larger food web

9. The producer in a food web is a/an
   (A) heterotroph
   (B) carnivore
   (C) autotroph
   (D) saprophyte
   (E) herbivore

10. Three environmental factors that can denature an enzyme are:
   (A) heavy metals, temperature, and substrate
   (B) substrate, temperature, and pH
   (C) pH, substrate, and heavy metals
   (D) heavy metals, temperature, and pH
   (E) water, air, and molecular oxygen

11. The structure that leads the egg to the oviduct is the
   (A) suspensory ligament
   (B) mature follicle
   (C) fimbria
   (D) cilium
   (E) corpus luteum

12. The incorrect statement related to the reason that chemical evolution is not occurring on earth today is
   (A) Appropriate energy sources such as strong ultraviolet radiation are not present today.

(B) Today's atmosphere is an oxidizing type which tends to break down rather than build up organic compounds.
(C) Any newly formed organic materials would be quickly used as a food source by existing organisms.
(D) Organic soup can currently only be formed in the region of the equator.
(E) The ozone layer of the atmosphere acts as an effective shield against ultraviolet radiation.

Questions 13 and 14 are based on the following graph.

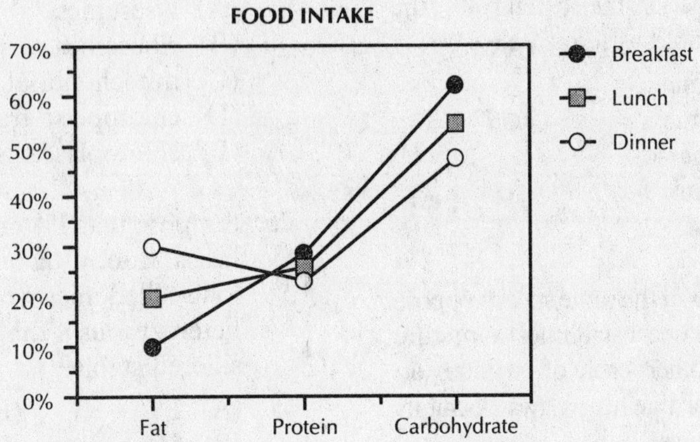

13. Based on the graph, which of the following statements about food intake is correct?
    (A) The average protein/fat ratio consumed was about 3/1.
    (B) The average fat/protein ratio consumed was about 3/1.
    (C) The average protein/carbohydrate ratio consumed was about 2/1.
    (D) The average protein/carbohydrate ratio consumed was about 1/2.
    (E) both (A and C)

14. From the data in the graph, which of the following conclusions is most reasonable?
    (A) Protein intake varies in direct proportion to carbohydrate intake.
    (B) Protein intake varies less than either carbohydrate or fat intake.
    (C) Protein intake varies in direct proportion to fat intake.
    (D) Fat and protein intake vary to a greater degree than does carbohydrate intake.
    (E) all of the above

15. The original experiments that showed it was possible for the gases of the primitive atmosphere to react with one another in the presence of external energy were performed by
    (A) Stanley Miller
    (B) Sidney Fox
    (C) Charles Darwin
    (D) Gregor Mendel
    (E) Louis Pasteur

16. Each of the major elements essential to life has cyclic processes that involve
    (A) a reservoir of the element, an exchange pool of producers and consumers that incorporate the element, and an abiotic community through which the element moves
    (B) photophosphorylation and fermentation
    (C) a minimum of six other elements
    (D) nitrifying and denitrifying bacteria
    (E) none of these

17. Vitamins often serve as parts of
    (A) complex carbohydrates like glycogen
    (B) large nonprotein groups called coenzymes
    (C) nucleic acids
    (D) inorganic molecules
    (E) simple carbohydrates like glucose

18. Cycles of activity or behavior that occur on a yearly basis are called
   (A) circadian rhythms
   (B) circannual rhythms
   (C) tidal rhythms
   (D) triannual rhythms
   (E) dodecannual rhythms

19. A mimic that lacks the defense of the organism it resembles is termed a
   (A) Mendelian mimic
   (B) Darwinian mimic
   (C) Mullarian mimic
   (D) Batesian mimic
   (E) none of these

20. When two animals of the same species oppose one another and one establishes a posture that makes the other back off without an attack, the animal that turns away exhibits a behavior known as
   (A) appeasement
   (B) dominance
   (C) aggression
   (D) altruism
   (E) none of these

21. The framework of all organic compounds is composed of straight chains, branched chains and rings of
   (A) nitrogen
   (B) carbon
   (C) hydrogen
   (D) oxygen
   (E) phosphorus

22. Systemic blood moves through the right side of the heart and directly into the
   (A) pulmonary vein
   (B) pulmonary artery
   (C) coronary artery
   (D) aortic arch
   (E) renal artery

23. The region of an enzyme where the substrate attaches is the
   (A) active site
   (B) carboxyl terminal end
   (C) amine terminal end
   (D) radicle group
   (E) none of these

24. It is currently believed that the specific enzymes necessary for the fixation of carbon dioxide into energy-rich glucose are located in the
   (A) lysosomes
   (B) ribosomes
   (C) mitochondrial cristae
   (D) chloroplast grana
   (E) chloroplast stroma

25. Suppose that 1% of a population of mice has a stubby tail. What percent of the long-tailed mice would be expected to be heterozygous if this population were considered stable?
   (A) 1%
   (B) 81%
   (C) 18%
   (D) 99%
   (E) 49%

26. The raw material for evolution is
   (A) mutation
   (B) gene flow
   (C) genetic drift
   (D) natural selection
   (E) none of these

27. The cell organelle characterized by stacked membranes called grana would most likely be involved in
   (A) cellular respiration
   (B) protein synthesis
   (C) photophosphorylation
   (D) intracellular digestion
   (E) heredity

28. Protein molecules that speed up chemical reactions have names that usually end in
   (A) -sis
   (B) -a
   (C) -ion
   (D) -ase
   (E) -phyte

29. The secondary structure of a protein is the
   (A) amino-acid sequence

(B) orientation of the amino-acid sequence
(C) three-dimensional structure of the oriented amino-acid sequence
(D) the structure that confers enzyme properties on the molecule
(E) double chain of nucleotides

30. The cell type that is not found in xylem is the
(A) vessel cell
(B) tracheid cell
(C) parenchyma cell
(D) endodermal cell
(E) none of these

31. The structures that are developed from the ectoderm are the
(A) liver and spleen
(B) stomach and small intestine
(C) brain and spinal chord
(D) trachea and lung
(E) heart and kidney

Question 32 is based on the following graph.

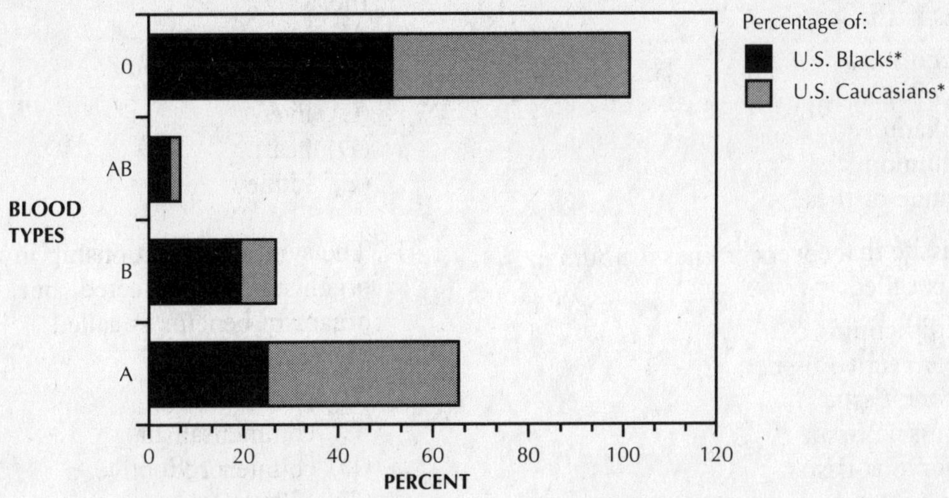

*** Information on other racial blood types not available.**

32. Based on the graph, which of the following statements about human blood types is correct?
(A) The frequency of all blood types varies between races.
(B) The frequency of blood type B varies between races.
(C) The frequency of blood types B and O varies between races.
(D) The frequency of blood type O is about the same in both races.
(E) both (B and D)

33. Directional selection in evolution
(A) tends to obliterate atypical phenotypes and thereby enhance population adaptations to a particular environment
(B) selects an atypical phenotype which is better adapted to a new environment than is a typical phenotype
(C) inevitably leads to the development of analogous structures
(D) results in the development of analogous structures from related structures in unrelated forms
(E) none of these

34. One of the best known examples of speciation is provided by the
(A) ciliary array in paramecium
(B) analogous structures found in nature
(C) the growth of bacteria on an agar medium
(D) existence of races among humans
(E) none of these

35. The research of Sidney Fox supports the idea that
    (A) amino-acid polymers were the first macromolecules to form on earth
    (B) nucleic acids are common to all cells
    (C) amino acids could be formed from simpler substances such as methane gas, ammonia gas, water vapor, and hydrogen gas
    (D) the first cells to form were autotrophic
    (E) photosynthesis came before cellular respiration

36. The embryonic sac that collects nitrogen wastes is the
    (A) chorion
    (B) yolk sac
    (C) allantois
    (D) amnion
    (E) none of these

37. The tissue that covers or lines the human body is called
    (A) epithelium
    (B) connective tissue
    (C) bone tissue
    (D) muscle tissue
    (E) nervous tissue

38. The most numerous type of tooth in the adult human is the
    (A) incisor
    (B) canine
    (C) premolar
    (D) molar
    (E) gingiva

39. The best explanation of how transport occurs in phloem is incorporated in the
    (A) pressure–flow theory
    (B) cohesion–tension theory
    (C) osmotic root–pressure theory
    (D) auxin–gibberellin theory
    (E) morphogenic–specialization theory

40. Deamination of amino acids occurs in the
    (A) pancreas
    (B) spleen
    (C) liver
    (D) heart
    (E) kidney

41. The symbiotic relationship in which one organism is unaffected but the other organism benefits is called
    (A) mutualism
    (B) parasitism
    (C) commensalism
    (D) counterproductive
    (E) succession

Questions 42 and 43 are based on the following graph.

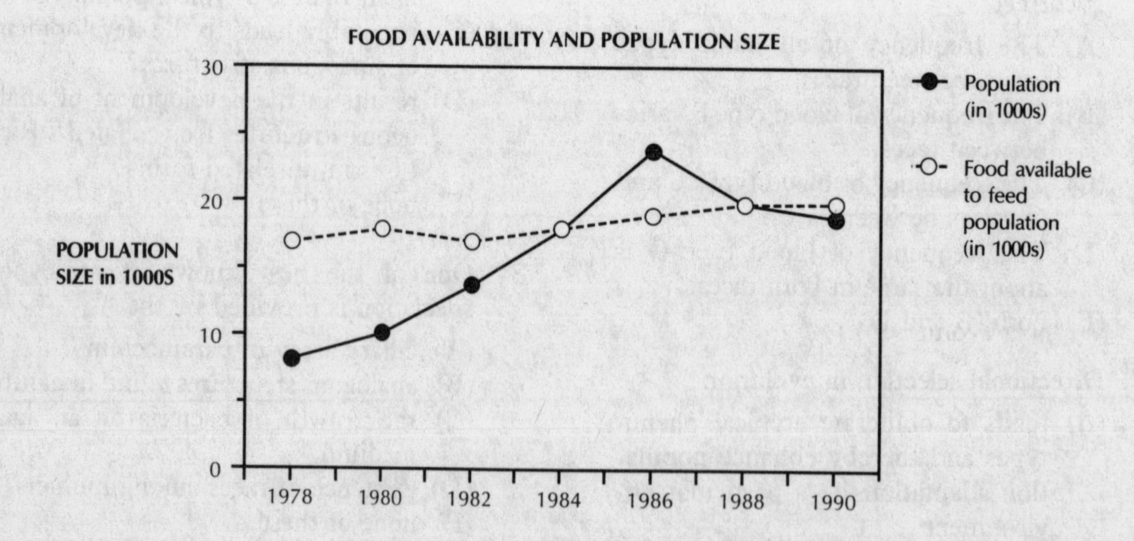

42. Based on the graph, which of the following statements about food availability and population size is correct?
    (A) Food availability has increased while population size has decreased.
    (B) Population size has increased while food availability has increased.
    (C) Population size has varied while food availability has slightly increased.
    (D) Food availability has increased while population has decreased.
    (E) Food availability has decreased while population size has remained the same.

43. From the data in the graph, which of the following conclusions is most reasonable?
    (A) 1986 was a peak year for food availability.
    (B) 1986 was the year for the smallest population size.
    (C) 1986 was a peak year for population size.
    (D) Food was more available in 1978 than it was in 1988.
    (E) Food availability in 1990 was the same as it was in 1978.

44. The correct sequence of the cell cycle is
    (A) interphase, prophase, telophase, metaphase, and anaphase
    (B) prophase, telophase, metaphase, interphase, and anaphase
    (C) prophase, metaphase, interphase, anaphase, and telophase
    (D) interphase, prophase, metaphase, anaphase, and telophase
    (E) interphase, telophase, prophase, anaphase, and metaphase

45. In which area is there no evidence to support evolution?
    (A) the fossil record
    (B) comparative embryology
    (C) comparative biochemistry
    (D) organic chemistry
    (E) comparative anatomy

46. The anabolic process in which sunlight, energy, and water are utilized and molecular oxygen is produced is
    (A) photosystem I
    (B) photosystem II
    (C) glycolysis
    (D) Krebs Cycle
    (E) respiratory chain

47. Which statement is untrue for the components of the nitrogen cycle?
    (A) Producers contribute nitrogen directly to the exchange pool.
    (B) Producers contribute nitrogen directly to consumers.
    (C) Decomposers contribute nitrogen directly to producers.
    (D) Decomposers contribute nitrogen directly to the exchange pool.
    (E) Consumers contribute directly to the nitrogen reservoir.

48. Bacteria involved in the nitrogen cycle are
    (A) decomposers
    (B) producers
    (C) consumers
    (D) carnivores
    (E) herbivores

49. The correct sequence of steps in urine formation are
    (A) tubular excretion, selective reabsorption, and pressure filtration
    (B) pressure filtration, tubular excretion, and selected reabsorption
    (C) pressure filtration, selective reabsorption, and tubular secretion
    (D) tubular excretion, pressure filtration, and selective reabsorption
    (E) tubular secretion, selective reabsorption, and pressure filtration

50. Double fertilization in plants refers to
    (A) two sperm nuclei combining with an egg nucleus resulting in a triploid zygote

(B) one sperm nucleus combining with the egg nucleus and the other sperm nucleus combining with the two polar nuclei
(C) the process by which two flowers exchange pollen
(D) the need for two pollen grains to unite with the antipodal cells
(E) none of these

51. The cycle that requires the involvement of several types of bacteria is the
(A) carbon cycle
(B) water cycle
(C) Calvin Cycle
(D) Krebs Cycle
(E) none of these

52. The bones of the forelimbs of bats, birds, cats, horses, humans, and whales can be considered
(A) analogous structures
(B) homologous structures
(C) clone structures
(D) unrelated structures
(E) vestigial structures

53. Scattered vascular bundles, leaves with parallel veins, and generally the lack of secondary growth all characterize most
(A) angiosperms
(B) bryophytes
(C) tracheophytes
(D) monocots
(E) dicots

54. The part of the flower in which the pollen tube forms is the
(A) anther
(B) pistil
(C) petal
(D) filament
(E) sepal

55. Which statement most likely describes the origin of life on earth?
(A) Autotrophs preceded heterotrophs.
(B) The primitive atmosphere of the earth was most likely a reducing type.
(C) The primitive atmosphere was most likely an oxidizing type.
(D) The first cells required molecular oxygen to produce energy.
(E) Cells have been formed in the laboratory in support of the origin of life on earth.

56. The male gametophyte in the flower is located in the
(A) ovary
(B) sepals
(C) receptacle
(D) anther
(E) stigma

57. *Thylakoid membrane* is to *light reaction* as *cristae membrane* is to
(A) glycolytic enzymes
(B) proteolytic enzymes
(C) Calvin Cycle enzymes
(D) respiratory chain enzymes
(E) photosystem II enzymes

58. The disappearance of the spindle, formation of nuclear membranes, and the reappearance of nucleoli characterize
(A) interphase
(B) prophase
(C) metaphase
(D) anaphase
(E) telophase

59. The process that converts a molecule of glucose into pyruvic acid with the net production of two molecules of ATP
(A) is the aerobic portion of cellular respiration
(B) includes enzymes of photosystems I and II
(C) is a form of anaerobic respiration
(D) is an anabolic form of metabolism
(E) occurs within the grana of the chloroplast

60. A chemical that is released by one organism and affects another organism of the same species is called a/an
    (A) allomone
    (B) hormone
    (C) pheromone
    (D) antigen
    (E) agglutinogen

61. Groups of unspecialized plant cells that are capable of cell division are known as
    (A) vascular tissues
    (B) meristematic tissues
    (C) epidermal tissues
    (D) parenchymal tissues
    (E) collenchymal tissues

62. The most numerous type of leucocyte is the
    (A) monocyte
    (B) lymphocyte
    (C) neutrophil
    (D) acidophil
    (E) basophil

63. The part of the developing embryo that becomes the primary root is the
    (A) cotyledon
    (B) hypocotyl
    (C) radicle
    (D) plumule
    (E) endosperm

64. The type of learning that employs a single type of stimulus time and time again until the response to the stimulus ceases is known as
    (A) conditioned learning
    (B) operant conditioning
    (C) habituation
    (D) insight learning
    (E) none of these

65. The theory of orderly development in higher animals is currently based upon the concept of
    (A) gastrulation
    (B) nondisjunction
    (C) transcription
    (D) translation
    (E) induction

66. Testosterone is produced by the adrenal cortex and by the
    (A) seminiferous tubule cells
    (B) interstitial cells
    (C) primary spermatocytes
    (D) epididymis cells
    (E) prostate cells

67. Enzymes derive their specificity from
    (A) their amino-acid sequence
    (B) the temperature and pH of the environment
    (C) the size and shape of the ribosomes on which they are synthesized
    (D) the substrate concentration
    (E) the size of the chromosome on which their templates are found

68. Nonenzymatic molecules that are close in shape to a true enzyme substrate but are nonreactant and therefore interfere with product formation are
    (A) competitive inhibitors
    (B) coenzymes
    (C) apoenzymes
    (D) competitive activators
    (E) pseudoenzymes

69. If a heterozygous for color blindness woman has a child by a man with normal vision, their chances of having a color-blind daughter are
    (A) 0%
    (B) 25%
    (C) 50%
    (D) 75%
    (E) 100%

70. Given the Hardy–Weinberg Law $p^2 + 2pq + q^2 = 1.0$, the $2pq$ represents the
    (A) homozygous dominant individual
    (B) heterozygous individual

(C) homozygous recessive individual
(D) rate of nutritional change in the population
(E) gene ratios in a multiple allelic series

71. The human karyotype is characterized by
(A) 22 pair of autosomes and one pair of sex chromosomes
(B) 46 autosomes and two sex chromosomes
(C) 23 pair of sex chromosomes
(D) 42 autosomes and 2 pair of sex chromosomes
(E) 23 autosomes and 2 sex chromosomes

72. In the human male, FSH is known to promote
(A) spermatogenesis in the seminiferous tubules
(B) the production of testosterone in the interstitial cells
(C) the production of estrogen in the seminiferous tubules
(D) contraction of the scrotum
(E) the secretion of seminal fluid

73. "The separation of genes when the gametes are formed with one gene of each allele being present in each gamete" is a modern statement of
(A) Mendel's Law of Segregation
(B) Mendel's Law of Independent Assortment
(C) the principle of incomplete and codominance
(D) Darwin's Theory of Natural Selection
(E) Lemark's Theory of Inherited Adaptations

74. Barnacles living on the mollusc shells, remoras living on sharks, and orchids living in the branches of trees are all examples of
(A) commensalism
(B) mutualism
(C) parasitism
(D) convergent evolution
(E) divergent evolution

75. The four human blood phenotypes in the ABO system result from the fact that
(A) Rh and ABO blood factors are incompatible
(B) the genes for A and B are codominant while the gene for O is recessive to both A and B
(C) O is the dominant blood type in the human population
(D) during evolution the Rh factor prevented other genes from expressing themselves
(E) each of the genes for all three human blood types must be on different chromosome pairs

Question 76 is based on the graph on page 205 which reflects the results of a nationwide study of antibiotic treatment of cows.

76. Based on the graph, which of the following statements about antibiotic treatment and mortality is correct?
(A) Mortality, due to infection, is directly proportional to antibiotic treatment.
(B) Mortality, due to infection, is inversely proportional to antibiotic treatment.
(C) There is no statistical relationship between treatment and mortality.
(D) Mortality increases as infection decreases.
(E) Increasing water intake decreases infection mortality.

77. The light sensitive pigment upon which flowering depends is
(A) florigen
(B) auxin
(C) phytochrome
(D) cytokinin
(E) gibberellin

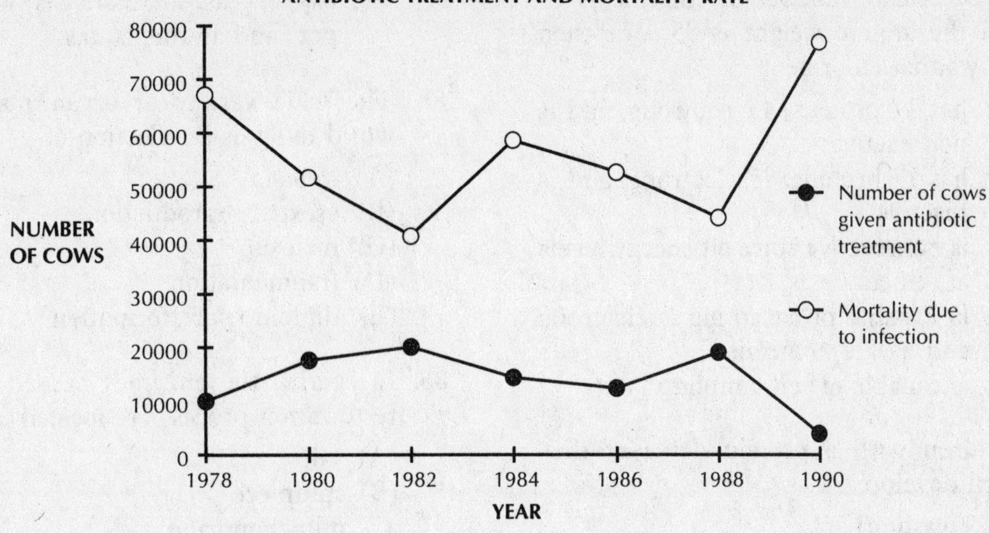

78. Marine molluscs and annelids share
    (A) larvae of the trochophore type
    (B) larvae of the dipeurula type
    (C) ventral nerve cord
    (D) closed circulation
    (E) the gastrovascular cavity

Questions 79 and 80 are based on the table below.

NUMBER OF CALORIES UTILIZED BY VARIOUS ACTIVITIES

KINDS OF ACTIVITY	CALORIES (PER HOUR)*
Walking up stairs	1100
Running (jogging)	570
Swimming	500
Vigorous exercise	450
Slow walking	200
Dressing and undressing	118
Sitting at rest	100

* includes Basal metabolic rate.

79. Based on the above table, which of the following statements about activity and calorie utilization is correct?
    (A) One-half hour of walking up stairs utilizes approximately the same number of calories as an hour of running.
    (B) One-half hour of running utilizes a greater number of calories than two hours of slow walking.
    (C) One-half hour of slow walking utilizes the same number of calories as two hours of sitting at rest.
    (D) both (A and B)
    (E) both (A and C)

80. From the data in the table, which of the following conclusions is most reasonable?
    (A) Exercise takes a great deal of motivation.
    (B) Slender people tend to exercise more.
    (C) It takes quite a bit of activity to burn off a high-calorie meal (1200 calories).
    (D) Both (B and C)
    (E) (A, B, and C)

81. The building blocks of triglycerides (neutral fats) are
    (A) glucose and sucrose
    (B) polypeptides and polynucleotides
    (C) starch and glycogen
    (D) glycerol and fatty acids
    (E) glucose and fructose

82. If the atomic number of chlorine is 17 and the atomic weight is 35, one then knows that chlorine
    (A) has 17 protons, 17 neutrons, and is nonreactive
    (B) has 17 protons, 17 electrons, and is reactive
    (C) is nonreactive since all energy levels are filled
    (D) is capable of accepting 2 electrons and has 17 neutrons
    (E) is capable of self-coupling

83. Children with a protein deficient diet often develop
    (A) kwashiorkor
    (B) Down's Syndrome
    (C) Turner's Syndrome
    (D) phlebitis
    (E) juvenile diabetes

84. One hypothesis about the origin of life on earth suggests that the primitive atmosphere was most likely composed of
    (A) hydrogen gas, oxygen gas, methane gas, and water vapor
    (B) hydrogen gas, ammonia vapor, methane gas, and water vapor
    (C) hydrogen gas, oxygen gas, methane gas, and ammonia gas
    (D) hydrogen gas, methane gas, water vapor, and ammonia gas
    (E) oxygen gas, methane gas, water vapor, and ammonia gas

85. The wide variety of organisms in the world today is a reflection of
    (A) mitosis
    (B) asexual reproduction
    (C) meiosis
    (D) fragmentation
    (E) diploid spore formation

86. In sperm, the enzymes needed for the fertilization process are located in the
    (A) tail
    (B) midpiece
    (C) mitochondrion
    (D) nucleus
    (E) acrosome

87. In the conversion of energy-poor carbon dioxide into energy-rich glucose
    (A) ATP is used
    (B) $NADPH_2$ is produced
    (C) ATP is produced
    (D) pyruvic acid is produced
    (E) molecular oxygen is produced

Questions 88 and 89. A scientist studying the influence of temperature and pH on enzyme reaction rate conducts a series of experiments. The results of her findings are reported in the line graph below.

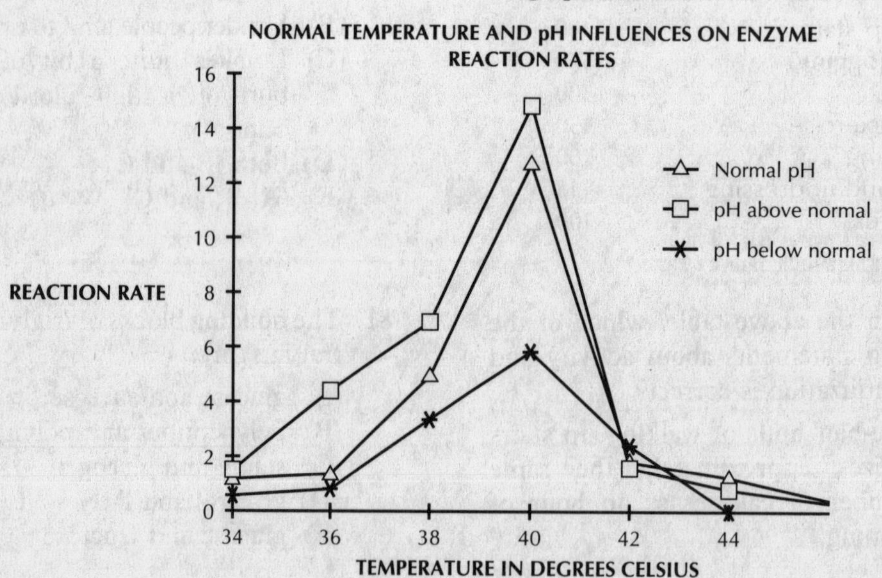

88. From the data plotted on page 206 it can be concluded that
    (A) enzymes at a pH below normal work more rapidly than do enzymes at a normal pH
    (B) enzymes at a pH above normal work more rapidly than do enzymes at a normal pH
    (C) enzymes at a pH below normal work more rapidly than do enzymes at a pH above normal
    (D) enzyme reaction rates double at a temperature above 42°C
    (E) normal enzyme reaction rates never exceed a level of 10

89. The most likely explanation for the fall in enzyme activity seen between 40°C and 42°C is
    (A) heat reduces the production of nucleic acids and therefore will interfere with lysosomal activity
    (B) enzymes above a temperature of 42°C become converted into fatty acids and glycerol
    (C) competition between enzymes and hormones becomes tipped in favor of the hormones
    (D) substrate feedback to the ribosomes stimulates the production of catalytic protein
    (E) under the influence of temperature enzyme rates increase until the enzymes become denatured

90. The mature egg is housed in the
    (A) primary follicle
    (B) secondary follicle
    (C) Graffian follicle
    (D) corpus luteum
    (E) seminal vesicle

91. A chemical or combination of chemicals that is capable of either accepting or donating hydrogen ions to solution are known as a/an
    (A) ion
    (B) atom
    (C) buffer
    (D) neutron
    (E) neuron

92. The green algae acetabularia was used in a series of experiments that demonstrated the importance of
    (A) chromosomes in heredity
    (B) nuclear control in the cell
    (C) the need for protein in the synthesis of hormones
    (D) the relationship between centrioles and spindle formation in mitosis
    (E) plant succession along beach communities

93. Hazardous wastes in the environment fall into three general categories:
    (A) heavy metals, inert gases, and chlorinated hydrocarbons
    (B) heavy metal, chlorinated hydrocarbons, and nuclear wastes
    (C) chlorinated hydrocarbons, nuclear wastes, and inert gases
    (D) chlorinated hydrocarbons, inert gases, and heavy metals
    (E) nuclear wastes, heavy metals, and inert gases

94. The two main steps in speciation are
    (A) geographic isolation of two populations followed by reproductive isolation
    (B) a change in the environment followed by natural selection
    (C) transcription and translation
    (D) mutation and communal living
    (E) convergent evolution followed by natural selection

95. Placental mammals belong to the subclass
    (A) Protheria
    (B) Matatheria
    (C) Eutheria
    (D) Vertebrata
    (E) Echinodermata

96. In a dihybrid cross the typical phenotypic ratio is
    (A) 3:1
    (B) 9:3:3:1
    (C) 2:1:2:1
    (D) 1:1
    (E) 3:3:3:3

97. The structure characterized by a pericycle and cortex is the
    (A) dicot leaf
    (B) monocot stem
    (C) older dicot stem
    (D) monocot root
    (E) dicot root

Directions: Each group of questions below consists of five lettered headings followed by a list of numbered phrases or sentences. For each numbered phrase of sentence, select the one heading that is most closely related to it. One heading may be used once, more than once, or not at all in each group.

Questions 98–101.

(A) carbon cycle
(B) nitrogen cycle
(C) tundra
(D) tropical rain forest
(E) ecological succession

98. The complete process is called sere.

99. The pioneer community is an integral part.

100. Associated with caribou, reindeer, and wolves in the summer and arctic foxes, snowshoe hares, and lemmings all year round.

101. Is particularly associated with photosynthesis and cellular respiration.

Questions 102–104.

(A) mitochondria
(B) ribosomes
(C) endoplasmic reticulum
(D) microtubules
(E) golgi apparatus

102. Thin cylinders composed of thirteen rows of tubulin, a globular protein arranged in a helical fashion.

103. The organelle which packages cellular secretions and which also synthesizes carbohydrates.

104. The organelle in which all of the enzymes for oxidative phosphorylation are located.

Questions 105–110.

(A) Class Osteichthyes
(B) Class Scyphozoa
(C) Class Insecta
(D) Class Hirudina
(E) Class Aves

105. The most numerous class of animals on the surface of the earth.

106. The class of animals that are aquatic and have a dorsal nerve cord.

107. Segmented worms that are ectoparasites.

108. Animals generally characterized as invertebrates with three pairs of jointed legs as adults.

109. The class of animals in which robins, eagles, and parrots are found.

110. Animals that are homeothermic and possess four-chambered hearts.

Questions 111–115. Use the following information to answer these questions.

A SEGMENT OF THE ORGANIC BASE SEQUENCE IN A PIECE OF DNA MOLECULE	tRNA CODE FOR THE FOLLOWING AMINO ACIDS
C G G T C C T G T G C A T G T	ALANINE (G C A) ARGENINE (C G G) SERINE (U C C) VALINE (G U U) CYSTINE (U G U)

111. The protein sequence for which this DNA is coded is
    (A) argenine–serine–cystine–alanine–cystine
    (B) argenine–serine–valine–alanine–cystine
    (C) serine–valine–argenine–alanine–cystine
    (D) valine–cystine–valine–arginine–valine
    (E) cystine–valine–serine–arginine–alanine

112. The mRNA transcribed from this DNA would have the base sequence:
    (A) **C G G T C C T G T G C A T G T**
    (B) **G C C A G G A C A C G U A G A**
    (C) **G C C U U A C G C U T T A T T**
    (D) **G C C U G G U C U C G U U C U**
    (E) none of these

113. The amino acid that is not coded for in this segment of DNA is
    (A) alanine
    (B) arginine
    (C) serine
    (D) valine
    (E) cystine

114. Assuming that the DNA segment is complete, a typical anticodon is
    (A) **T C C**
    (B) **U G G**
    (C) **G C A**
    (D) **A T T**
    (E) **G G G**

115. Assuming that the DNA segment is complete, a typical codon is
    (A) **A A G**
    (B) **C A A**
    (C) **C C A**
    (D) **C A C**
    (E) **T C C**

GO ON TO THE NEXT PAGE.

Questions 116–120.

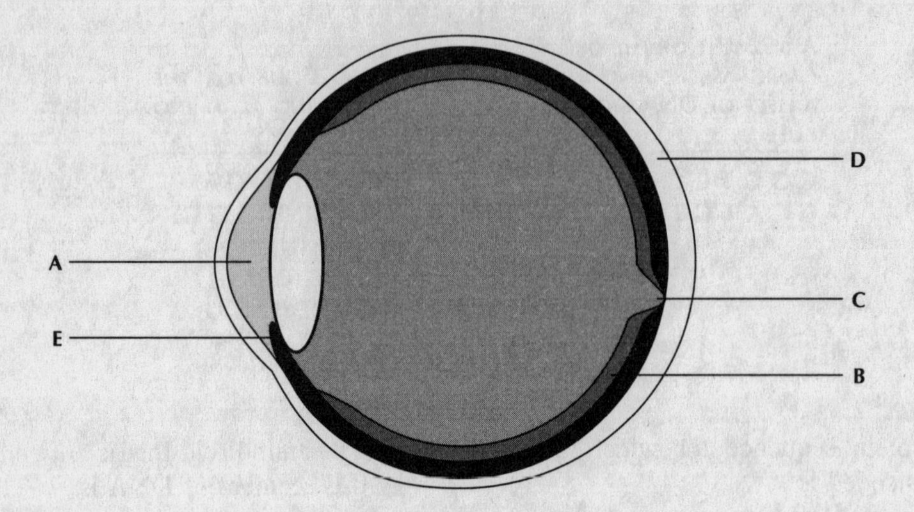

116. The layer in which there are numerous neurons.

117. The structure in which one finds an alkaline watery fluid.

118. The structure bounded by the cornea on one side and the iris and lens on the other side.

119. The region in which there are only cone cells and vision is most acute in daylight.

120. The tough white fibrous connective tissue layer that is modified as the cornea anteriorly.

SECTION 2

Time—90 minutes

Percent of Total Grade—40

The questions in this section are mandatory. Answer *ALL* four questions.

1. All living things face the same problems of life: gas exchange, food acquisition, nitrogen waste elimination, transport, etc. Different organisms have come up with different solutions to these problems.

a. List and describe the four main structures of gas exchange found in animals. Give an example for each type listed.

b. Briefly discuss the structure of the heart in:

(i) fishes
(ii) amphibians
(iii) reptiles
(iv) birds and mammals

2. Diagram and label each of the following:

a. typical cross section of a dicot root
b. typical cross section of a monocot leaf
c. typical cross section through the tubular portion of the alimentary tract
d. typical longitudinal section through a nephron
e. typical longitudinal section through a flower

3. Modern genetics is somewhat different from Mendelian genetics.

a. For each of the following examples identify the type of genetic phenomenon that is involved.

(i) short-legged "Dexter" cattle in the homozygous recessive condition
(ii) skin color in humans
(iii) Down's syndrome
(iv) ABO blood type in humans
(v) color blindness in humans
(vi) O blood type in humans
(vii) a phenotypic ratio of 2:1
(viii) calico (tortoise shell) cats
(ix) A and B together in human blood types
(x) hemophilia

b. Summarize Mendel's conclusions in modern terms.

4. The cell is a unit of biological activity that is delimited by a semipermeable membrane and is capable of self-reproduction in a medium free of other living systems.

> "Plastids and mitochondria were originally free-living prokaryotes that early in evolution took up residence inside of eukaryotic cells.

a. Describe the structure of both the chloroplast and the mitochondrion. State the function of each in the cell with particular reference to what goes on where.

b. Describe the evidence that supports the quote above.

ANSWER KEY

SECTION 1

1. **B**	25. **C**	49. **C**	73. **A**	97. **E**
2. **E**	26. **A**	50. **B**	74. **A**	98. **E**
3. **A**	27. **C**	51. **E**	75. **B**	99. **E**
4. **A**	28. **D**	52. **B**	76. **B**	100. **C**
5. **C**	29. **B**	53. **D**	77. **C**	101. **A**
6. **B**	30. **D**	54. **B**	78. **A**	102. **D**
7. **B**	31. **C**	55. **B**	79. **E**	103. **E**
8. **A**	32. **E**	56. **D**	80. **C**	104. **A**
9. **C**	33. **B**	57. **D**	81. **D**	105. **C**
10. **D**	34. **E**	58. **E**	82. **B**	106. **A**
11. **C**	35. **A**	59. **C**	83. **A**	107. **D**
12. **D**	36. **C**	60. **C**	84. **D**	108. **C**
13. **D**	37. **A**	61. **B**	85. **C**	109. **E**
14. **B**	38. **D**	62. **C**	86. **E**	110. **E**
15. **A**	39. **A**	63. **C**	87. **A**	111. **A**
16. **A**	40. **C**	64. **C**	88. **B**	112. **B**
17. **B**	41. **C**	65. **E**	89. **E**	113. **D**
18. **B**	42. **C**	66. **B**	90. **C**	114. **B**
19. **D**	43. **C**	67. **A**	91. **C**	115. **E**
20. **A**	44. **D**	68. **A**	92. **B**	116. **B**
21. **B**	45. **D**	69. **A**	93. **B**	117. **A**
22. **B**	46. **B**	70. **B**	94. **A**	118. **A**
23. **A**	47. **E**	71. **A**	95. **C**	119. **C**
24. **E**	48. **A**	72. **A**	96. **B**	120. **D**

DIAGNOSTIC CHART

SECTION 1

This chart will provide you with the opportunity of identifying those areas of biology in which you have done well and those areas in biology in which you need to improve your knowledge and understanding. For each correct answer, place an X in the space provided. You are doing well if you have 80–90% of the answers correct in each topic area. You are also doing well if your *total* score is 85% or better.

TOPIC	QUESTIONS ON THE EXAMINATION	NUMBER CORRECT
Basic Chemistry	1 21 29 81 82 91	______
Cell Structure and Function	27 55 92 102 103 104	______
Enzyme Structure and Function	10 17 23 28 67 68	______
Cell Division	5 14 58 71 85	______
Energy Transforms in Living Systems	19 24 46 56 59 87	______
Origin of Life	12 15 35 55 84	______
Taxonomy	78 95 105 116 107 108 109 110	______
Higher Plants: Structure and Function	30 39 53 61 77 97	______
Higher Plants: Reproduction and Development	2 7 50 54 56 63	______
Higher Animals: Structure and Function	3 22 37 38 40 49 62 67 83 116 117 118 119 120	______
Higher Animals: Reproduction and Development	4 11 31 36 65 66 72 86 90	______
Heredity	69 73 75 96 111 112 113 114 115	______
Evolution	25 26 33 34 45 52 70 94	______

Topic	Questions	Number Correct
Ecology	8 9 16 20 47 48 51 74 93 98 99 100 101	______
Behavior	18 20 41 60 64	______
Graph, Chart, or Table Interpretation	6 13 14 32 42 43 76 79 80 88 89	______
	Total Correct	______

EXPLANATORY ANSWERS

SECTION 1

1. **(B)** is correct. (A) Both are disaccharides. (C) Both are polysaccharides. (D) The first is a monosaccharide while the second is a disaccharide. (E) Both are disaccharides.

2. **(E)** is correct. The fruit of a plant is the part that bears the seeds.

3. **(A)** is correct. (B) are foreign proteins (in some cases, polysaccharides). (C) are proteins that maintain blood osmotic pressure, pH, blood clotting, and fight infection. (D) are substances containing attenuated bacteria and viruses. (E) is the amount of antibody in the circulation after vaccination.

4. **(A)** is correct. Nuclear transplant experiments in frogs and mice have clearly demonstrated that each cell of the body carries the full range of genetic information for the individual. Cell specialization is the result of the expression of a specific set of genes within the individual's gene pool.

5. **(C)** is correct.

6. **(B)** is correct.

7. **(B)** is correct. (A, C, and D) are typically composed of diploid cells. (E) is composed of several haploid nuclei.

8. **(A)** is correct. Nonrenewable fossil fuel energy is used inefficiently, material resources enter the system and do not cycle, and outputs of the system produce a wide variety of pollutants.

9. **(C)** is correct. Produces manufactured complex compounds from simple compounds such as atmospheric carbon dioxide and environmental water, and sunlight energy.

10. **(D)** is correct.

11. **(C)** is correct. (A) hold the ovary in place. (B) is a structure in the ovary. (D) are in the oviducts. They provide the mechanism for the movement of the egg along the oviduct. (E) are the gland-like remains of the Graffian follicle after the egg has been released.

12. **(D)** is the untrue statement.

13. **(D)** is correct.

14. **(B)** is correct. Protein varies about 7–8% while carbohydrates vary between 15–20% and fats vary between 20–25%.

15. **(A)** is correct. (B) conducted the experiments that supported the hypothesis that amino acid polymers were the first macromolecules to develop during the origin of life on earth. (C, D, and E) are not known for speculations or experiments related to the origin of life.

16. **(A)** is correct.

17. **(B)** is correct.

18. **(B)** is correct. (A) occurs on a daily basis. (C) occurs approximately twice a day. (D) would be a cycle that occurs three times a year but no such cycle is known in nature. (E) would be a cycle that occurs twelve times a year but no such cycle is known in nature.

19. **(D)** is correct. (A and B) are nonsense. (C) is a mimic that possesses the same defense and similar appearance to that of the organism it is mimicking.

20. **(A)** is correct. (B) is the behavior of the other animal. (C) is the behavior exhibited before the appeasement behavior. (D) is a behavior of one animal that increases the fitness of another animal rather than the animal performing the behavior.

21. **(B)** is capable of forming straight chains, branched chains, and rings that all serve as the framework for carbohydrates, lipids, proteins, and nucleic acids. (A, C, D, and E) are

all important elements that are found attached to the carbon skeletons.

22. **(B)** is correct. Systemic blood enters the right atrium, moves through the tricuspid valve into the right ventricle and then passes through the right semilunar valve into the pulmonary artery.

23. **(A)** is the site in which a substrate becomes properly oriented for the enzyme reaction.

24. **(E)** is correct. (A) are structures that contain powerful proteolytic enzymes used for intracellular digestion. (B) are structures involved in protein synthesis. (C) are structures that contain the enzymes used in the respiratory chain. (D) are structures that contain enzymes used in photosystems I and II.

25. **(C)** is correct. Applying the Hardy–Weinberg Law we find: $q^2 = 0.01$ and $q = 0.1$. Since $p + q = 1.0$, then $p = 0.9$. If this is the case then $p^2 = 0.81$. The heterozygote is $2pq$ which is equal to 0.18. (A) is the expected homozygous recessive gene frequency. (B) is the expected homozygous dominant gene frequency. (D and E) are nonsense.

26. **(A)** is correct. (B) results from the introduction of new genes into a gene pool of a particular population. These new genes are usually introduced by members of another population of that species. (C) results in a reduction in gene pool variations. (D) works on the mutations in the species by filtering out those mutations poorly adapted for the environment.

27. **(C)** is correct. Chloroplasts contain grana which incorporate enzymes needed for photophosphorylation. (A) is the mitochondrion. (B) is the ribosome. (D) is the lysosome. (E) is the chromosome.

28. **(D)** is correct. (A) is a typical ending meaning "the process of." (B) is a typical plural ending. (C) is a typical singular ending. (E) is a typical ending meaning "plant."

29. **(B)** is correct. (A) is the primary structure of a protein. (C and D) is the tertiary structure of a protein. (E) is the structure of DNA.

30. **(D)** are cells typical of the outside of the stele in the root. Many endodermal cells have bands of fatty material. These bands, the Casparian strips, prevent substances from moving between them.

31. **(C)** is correct. (A, B, D, and E) are all derived from mesoderm and endoderm.

32. **(E)** is correct.

33. **(B)** is correct. (A) is stabilizing selection. (C) is nonsense. (D) is untrue as analogous structures arise from totally unrelated forms that have no similar structures.

34. **(E)** is correct. The finches of the Galápagos Islands are an example of speciation. They are believed to have descended from mainland finches. Several different species have evolved to fill previously unfilled or poorly filled niches.

35. **(A)** is correct. Fox was able to demonstrate that mixtures of amino acids will combine in a preferred order when exposed to dry heat.

36. **(C)** is correct. (A) is the membrane sac that carries on gas exchange. (B) is the sac that surrounds the remaining yolk material. (D) is the sac that contains the amniotic fluid that bathes and protects the developing embryo.

37. **(A)** forms a continuous layer or sheet over the entire surface of the body and forms the lining of body cavities. (B) binds structures, provides protection, serves as a support framework, serves as a transport mechanism, fills spaces, and stores energy. (C) is a form of connective tissue. (D) provides for movement. (E) conducts information to and away from a central coordination center.

38. **(D)** There are twelve molars. (A) There are eight incisors. (B) There are four canines. (C) There are eight premolars. (E) is the soft tissue of the gum.

39. **(A)** is correct. (B) is the best explanation for xylem transport in tall plants. (C) is another possible explanation for xylem transport in short plants. (D and E) are nonsense.

40. **(C)** is correct. The removed amine group is first converted into ammonia and then into a less toxic nitrogen compound, urea.

41. **(C)** is correct. (A) is the situation where both organisms benefit. (B) is the situation in which one organism benefits while the other organism is harmed. (D) is nonsense. (E) is an ecological succession.

42. **(C)** is correct. From 1978 to 1990, there is a slight trend upward in the food availability but the population trend is upward from 1978 to 1986 and downward from 1986 to 1990.

43. **(C)** is correct.

44. **(D** is correct.

45. **(D)** is correct. Organic chemistry basically deals with the synthesis of organic compounds, many of which do not exist in nature.

46. **(B)** is correct. (A) utilizes only sunlight energy and uses this energy in the production of ATP. (C) is the process that converts glucose into pyruvic acid with the net production of a small quantity of ATP. (D) is the process that converts pyruvic acid to carbon dioxide with the release of energy needed for the respiratory chain production of ATP. (E) is the process that utilizes energy released by the Krebs Cycle in order to produce ATP in the presence of molecular oxygen.

47. **(E)** is correct. See diagram below.

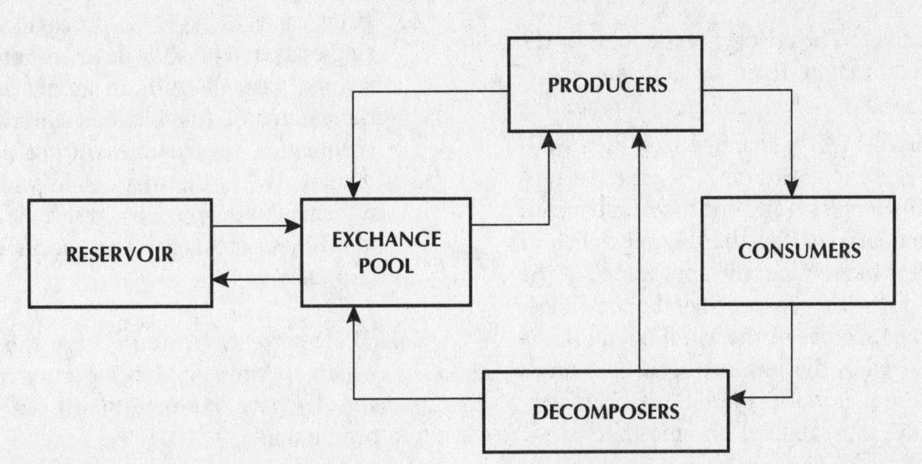

48. **(A)** is correct. (B) represents the plants. (C) represents the animals that consume plants and/or animals. (D) represents secondary consumers. (E) represents the producers.

49. **(C)** is correct. Pressure filtration occurs in the Bowman's capsule. Selective reabsorption occurs primarily in the proximal convoluted tubules. Tubular excretion occurs primarily in the distal convoluted tubules.

50. **(B)** is correct.

51. **(E)** is correct. The nitrogen cycle requires nitrogen-fixing bacteria to capture atmospheric nitrogen and bind it to organic compounds. Nitrifying bacteria convert ammonia into nitrites and nitrates. Denitrifying bacteria convert ammonia and nitrates back into atmospheric nitrogen.

52. **(B)** is correct. They all possess similarities in bond configuration. (A) have similar function by very different structures. (C) is nonsense. (D) is incorrect as they are unrelated structures. (E) are generally poorly developed, seemingly useless structures in one organism that in a different species are fully developed and functional.

53. **(D)** is correct. (A) includes monocots and dicots. (B) includes the first land plants. They lack any vascular tissues. (C) includes ferns, gymnosperms, and angiosperms. (E) includes plants with organized vascular bundles, leaves with net veins, and generally exhibit secondary growth.

54. **(B)** is correct. (A) is that part of the flower that produces the pollen. (C) is one of the whorls of modified leaves that are usually

brightly colored. (D) is the slender stalk that supports the anther. (E) is the whorl of modified leaves outside of the whorl of petals.

55. **(B)** is correct. As the earth cooled, the lightest atoms reacted with one another. Since hydrogen was the lightest and most abundant element, it is likely that the atmosphere was a reducing type.

56. **(D)** is correct. (A) is the structure which contains the female gametophyte. (B) are green, modified leaves that form a whorl around the petals. (C) is the basal tissue around the ovary. (E) is an enlarged, sticky knob at the tip of the pistal.

57. **(D)** is correct. The cristae are the folds of the inner membrane of the mitochodrion.

58. **(E)** is correct. (A) is characterized by a well-defined nuclear membrane, a granular appearing nucleoplasm and one to several nuclei. (B) is characterized by the disappearance of the nuclear membrane, the appearance of the chromosomes, the separation of the centrioles, and the appearance of the astral fibers. (C) is characterized by the lining up of chromosomes along the equatorial plane. (D) is characterized by the separation of chromosome pairs.

59. **(C)** is glycolysis. (A) is the Krebs Cycle and respiratory chain. (B) The photosystems I and II are the light reactions of photosynthesis. (D) Glycolysis as well as the rest of cellular respiration is catabolic metabolism. (E) Chloroplasts carry out the anabolic metabolism known as photosynthesis.

60. **(C)** is correct. (A) is a substance released by one organism and affects an organism of a different species. (B) is a substance released by one part of the body of an organism and affects a different part of the body of that organism. (D) is a foreign protein that usually triggers an immune response in the body of the recipient. (E) is a substance that causes erythrocytes to stick together.

61. **(B)** is correct. (A,C,D, and E) are all specialized plant tissues.

62. **(C)** is correct. Neutrophils normally represent 60–70% of all leucocytes.

63. **(C)** is correct. (A) is the seed leaf of the plant. (B) is the "embryonic stem." (D) is the "embryonic leaf." (E) is the tissue that stores foods in the form of carbohydrates.

64. **(C)** is correct. (A) results from an association of an irrelevant stimulus with a particular behavior. (B) is also called "trial and error" learning—a learning resulting from either a reward or punishment for selecting one of several alternative behaviors. (D) is the behavior that allows the animal to solve a problem by using previous experience.

65. **(E)** is correct. (A) is the process by which the single-layered blastula develops into the three-layered gastrula with an archenteron. (B) is the failure of homologous chromosomes or chromatids to separate during gamete formation. (C) is the manner in which mRNA is formed in the nucleus. (D) is the conversion of mRNA's message into a specific amino acid sequence.

66. **(B)** is correct. (A and C) are the cells from which sperm cells develop after meiosis. (D and E) have nothing to do with hormone production.

67. **(A)** is correct. The amino-acid sequence will determine the polypeptide chain orientation and ultimately, the three-dimensional shape of the protein.

68. **(A)** is correct. (B) are nonprotein groups that work with apoenzymes, the protein groups. (D and E) are nonsense.

69. **(A)** is correct. See the diagram on page 219. The gene locus for color blindness is on the X-chromosome. There is no corresponding gene locus on the Y-chromosome. For sons there is a 50% chance of having a child who is color blind.

70. **(B)** is correct. (A) is p^2. (C) is q^2. (D) does not apply since the law depends on a population in which no mutation occurs. Once there is mutation, gene frequency changes and evolution is considered to have occurred.

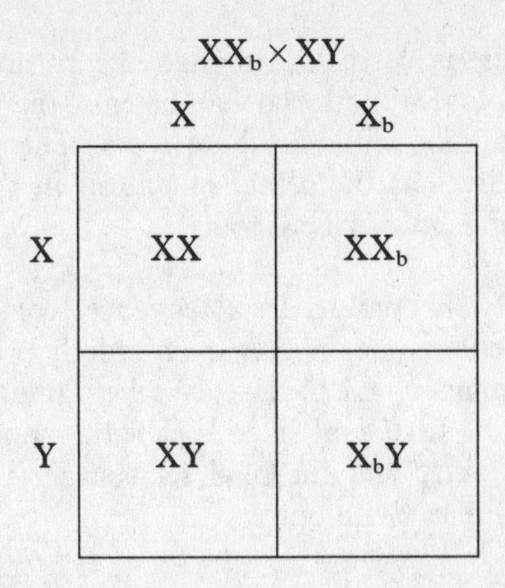

71. **(A)** is correct.

72. **(A)** is correct. (B) is under the influence of LH. (C,D, and E) are either nonsense or do not apply.

73. **(A)** is correct. (B) states that pairs of factors separate from one another to form gametes with many different gene combinations. (C) has to do with relationships between or among alleles. (D and E), although related, deal with evolution, not heredity.

74. **(A)** is correct. This is the form of symbiosis in which one organism benefits while the other organism does not, but is unharmed.

75. **(B)** is correct. Since the genes for ABO blood types appear at the same locus for a particular chromosome pair, only two alleles can be present at any time. The genes for A and B are codominant while the gene for 0 is recessive. The result is six genotypes that yield four phenotypes.

A PHENOTYPE = AA OR AO GENOTYPES
B PHENOTYPE = BB OR BO GENOTYPES
AB PHENOTYPE = AB GENOTYPE
O PHENOTYPE = OO GENOTYPE

76. **(B)** is correct. When one is on the rise, the other is on the decline and visa versa.

77. **(C)** is the pigment that absorbs red light of a particular wavelength. (A,B,D, and E) are all plant hormones.

78. **(A)** is correct. (B) is the larval type shared by echinoderms and certain chordates. (C) is nonsense. (D) Both have open circulatory systems. (E) The gastrovascular cavity is common to cnidarians and platyhelminths.

79. **(E)** is correct. (B) According to the table, one-half hour of running will require 285 calories while two hours of slow walking requires 400 calories.

80. **(C)** is correct. (A and B) cannot be determined from the available information.

81. **(D)** is correct. (A) The first is a monosaccharide while the second is a disaccharide. (B) Polypeptides are chains of amino acids while polynucleotides are chains of nucleotides. (C) Both are polysaccharides. (D) Both are six carbon monosaccharides.

82. **(B)** is correct. The atomic number is the number of protons in the nucleus of the atom. Atoms are electrically stable, therefore the number of protons equals the number of electrons. The atomic weight is the sum of the protons and neutrons in the nucleus of the atom. The number of neutrons can be determined by subtracting the atomic number from the atomic weight. Atomic reactivity depends on the ability of the atom to share, donate, or accept electrons.

83. **(A)** is correct. (B and C) are genetic disorders. (D) is the inflammation of a vein. (E) is a genetic metabolic disorder.

84. **(D)** is correct. It is believed that hydrogen was the most abundant element in the early atmosphere of the earth. Hydrogen readily combined with carbon, oxygen, and nitrogen to form gases.

85. **(C)** provides the means for the recombination of genetic materials and thereby results in new sets of chromosomes necessary for the variety of life.

86. **(E)** is correct. (A) contains the 9+2 pattern of microtubules typically found in cilia and flagella. (B) contains numerous mitochondria needed to provide the energy for flagellum whipping. (C) are energy producing structures found in the midpiece. (D) is found in the head of the sperm.

87. **(A)** is correct. (B) is used and not produced. (C) is used and not produced. (D) is part of cellular respiration. (E) is produced by photolysis of water in photosystem II.

88. **(B)** is correct.

89. **(E)** is correct. (A,B,C, and D) are nonsense.

90. **(C)** is correct. (A and B) are stages of egg and follicle development that lead to the mature Graffian follicle. (D) is the gland-like structure that develops from the Graffian follicle after ovulation occurs. (E) is a gland that contributes secretions to those of the prostate gland and the Cowper's gland.

91. **(C)** is correct. (A) is a charged atom or molecule. (B) is the smallest particle that is indivisible by ordinary chemical means. (D) is an atomic particle that carries no electrical charge. (E) is the functional cell of the nervous system.

92. **(B)** is correct. Acetabularia is a very large single-celled organism with a base, stalk, and cap. It is useful in cross-species grafts.

93. **(B)** is correct. Inert gases do not react chemically and therefore present no significant environmental threat.

94. **(A)** is correct. Geographic isolation allows for divergent evolution for three reasons. (1) Differences in gene pools of the isolated populations. (2) Separated gene pools will each be subject to different mutations and gene recombinations. (3) As each population is in a different environment, each is subject to different selective pressures.

95. **(C)** is correct. (A) is the subclass of egg-laying mammals. (B) is the subclass of pouched mammals. (D) is the subphylum of chordates which are characterized by a spinal cord. (E) is the phylum of starfish, sea cucumbers, and sand dollars.

96. **(B)** is correct. A dihybrid cross involves two different sets of alleles.

97. **(E)** is correct. Although the cortex can be found in both roots and stems, the pericycle is only found in dicot roots. The pericycle serves as the center of mitosis in the development of lateral roots.

98. **(E)** is correct. In a sequence, communities replace one another in an orderly predictable manner. Each stage is called a seral stage. The first seral stage is called a pioneer community and the final seral stage is called a climax community.

99. **(E)** is correct. The first seral stage of ecological succession is the pioneer community. These are usually plants capable of surviving harsh conditions. This seral stage paves the way for other communities.

100. **(C)** is correct. Tundra is a region of permafrost, a layer of subsoil which never unfreezes.

101. **(A)** is correct. Carbon dioxide of respiration is released into the atmosphere. Decomposition of dead organisms also releases carbon dioxide. Photosynthesis fixes energy-poor carbon dioxide into energy-rich glucose.

102. **(D)** is correct. Microtubules exist independently in the cytoplasm and are found in cilia, flagella, and centrioles. They are rapidly assembled and disassembled allowing for changes in cell structure and shape.

103. **(E)** is correct. This stack of saccules and associated vacuoles is specifically arranged with respect to the cell nucleus.

104. **(A)** is correct. Oxidative phosphorylation occurs in the electron transport system. The enzymes for this process are located in an assembly-line fashion along the cristae of the mitochondria.

105. **(C)** is correct. It is estimated that there are 500,000 to 750,000 species of insect on the earth today. They all belong to the Class Insecta of the Phylum Arthropoda.

106. **(A)** are the fishes. They are members of the Subphylum Vertebrata. They typically ex-

hibit the dorsal nerve cord. Nerve cords, where they exist among the invertebrates, tend to be ventral structures.

107. (**D**) is correct. These segmented worms are the leeches. They feed by attaching themselves to a host, inflicting a wound and drawing the blood of the host for nourishment. They belong to the Phylum Annelida.

108. (**C**) is correct. Although some larval forms of insect do not have three pair of legs, most adult insects do. They are invertebrates that belong to the Phylum Arthropoda.

109. (**E**) These are the birds. They are characterized by the specialized structure called the feather. Both the aves and mammals are homeotherms, animals that can maintain a body temperature above that of the environment. Both have four-chambered hearts. They both belong to the Phylum Chordata.

110. (**E**) is correct. See answer 109.

111. (**A**) is correct. DNA makes complementary mRNA. The thymine in DNA is replaced with uracil in RNA. The codes on tRNA are complementary to the code on the mRNA. It is the specific coding that brings the correct amino acid into position during protein synthesis.

112. (**B**) is correct. mRNA is a complement of the DNA with all thymines being replaced by uracil.

113. (**D**) is correct. All of the other choices have codes that make the DNA code with only one change. Uracil replaces thymine.

114. (**B**) is correct. The codon is the organic base triplet in DNA while the anticodon is the complementary organic base triplet in mRNA.

115. (**E**) is correct.

116. (**B**) is the retina. The innermost layer of the eye is composed of three sublayers with the innermost sublayer being composed of specialized epithelial cells, the rods and cones. The middle sublayer is composed of bipolar neurons. The outermost sublayer is composed of ganglion cells.

117. (**A**) is the anterior chamber which is filled with an alkaline fluid, aqueous humor.

118. (**A**) See answer 117.

119. (**C**) is the fovea centralis, an oval depression where there are usually only cone cells.

120. (**D**) is the sclera. It becomes modified anteriorly to become the transparent cornea.

MODEL ESSAYS

SECTION 2

Because subjectivity enters into the grading of any essay, it is not possible to provide *the* answer to a particular essay question. When you compare your answers with the model answers that follow, keep in mind that there are a number of ways in which they could have been written. Pay attention to the content and the organization of the answer; see whether you have followed a similar style for content and organization.

Question 1

a. (i) The body surface or skin in lower forms like earthworms serves as a surface of gas exchange. This mechanism is effective only if the animal has a relatively high surface area to volume ratio. The surface must always be kept moist for effective gas exchange. This mechanism requires a circulatory system to transport gases in the body.
(ii) The tracheal systems of insects, centipeds, millipeds, and some spiders are branched air tubes that extend throughout the body. The surface openings of this tubular system are called spiricles. Trachial systems do not require a circulatory system for gas transport. As with all gas-exchange surfaces, trachial tubes must be kept moist.
(iii) The gills of fishes are feathery tissue outgrowths that exchange gases with water. Gills require a circulatory system for gas exchange and transport around the body. There is no problem with keeping the gas-exchange surfaces wet in fishes.
(iv) The lungs of amphibians, reptiles, birds, and mammals are continuously branching trees that lead to a specific region of the body. The terminal branches end in saclike alveoli. The alveoli are highly vascularized, as they are the regions of gas exchange.

b. (i) The fish heart is essentially a two-chambered structure through which the blood travels only once before going to the gas-exchange surface of the body. A sinus venosus leads the blood to a single atrium which leads blood in turn into a single ventricle. Blood from the ventricle moves blood into the conus arteriosus and on into the dorsal aorta.
(ii) The amphibian heart is a three-chambered structure through which blood travels twice; once to the lungs and a second time to the body. There are two separate atria but only a single ventricle in which oxygenated blood and deoxygenated blood tend to mix.
(iii) The reptilian heart is a three-chambered structure similar to that of the amphibian heart. The one big difference is the partially

divided ventricle in reptiles, which tends to reduce the amount of deoxygenated and oxygenated blood mixing.

(iv) The bird and mammalian hearts are four-chambered structures. There are two completely separate atria and two completely separate ventricles. The right side of the heart collects deoxygenated blood from the body and delivers it to the lungs, while the left side of the heart collects oxygenated blood from the lungs and delivers it to the body.

Question 2

Fig. a. Typical cross section of a dicot root.

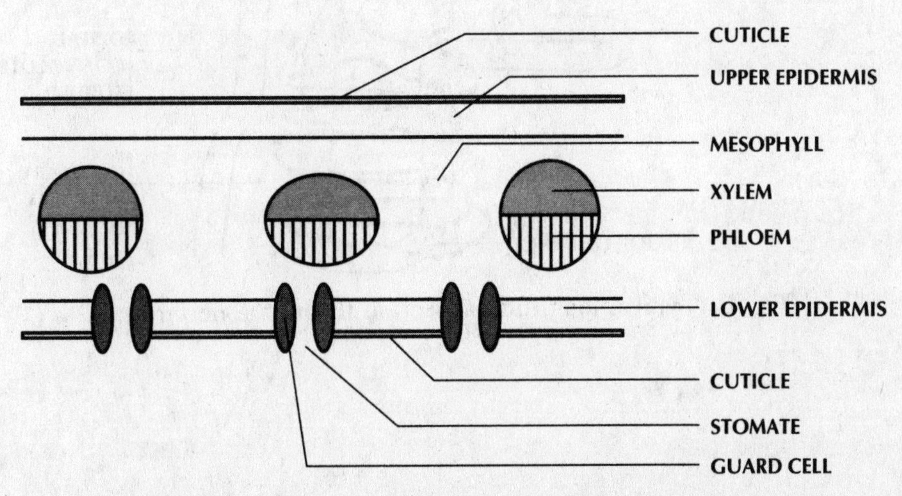

Fig. b. Typical cross section of a monocot leaf.

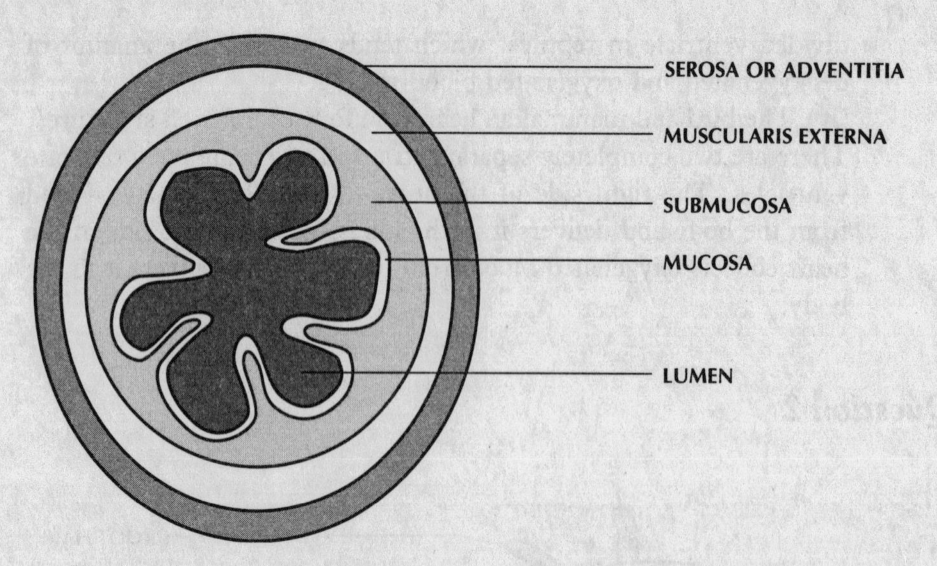

Fig. c. Typical cross section through the tubular portion of the alimentary tract.

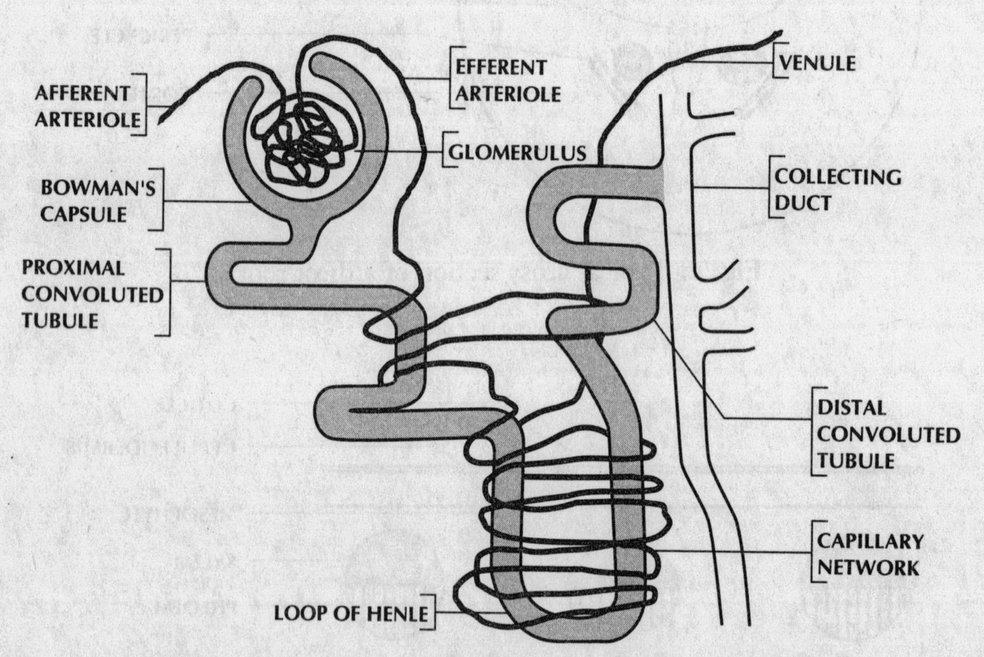

Fig. d. Typical longitudinal section through a nephron.

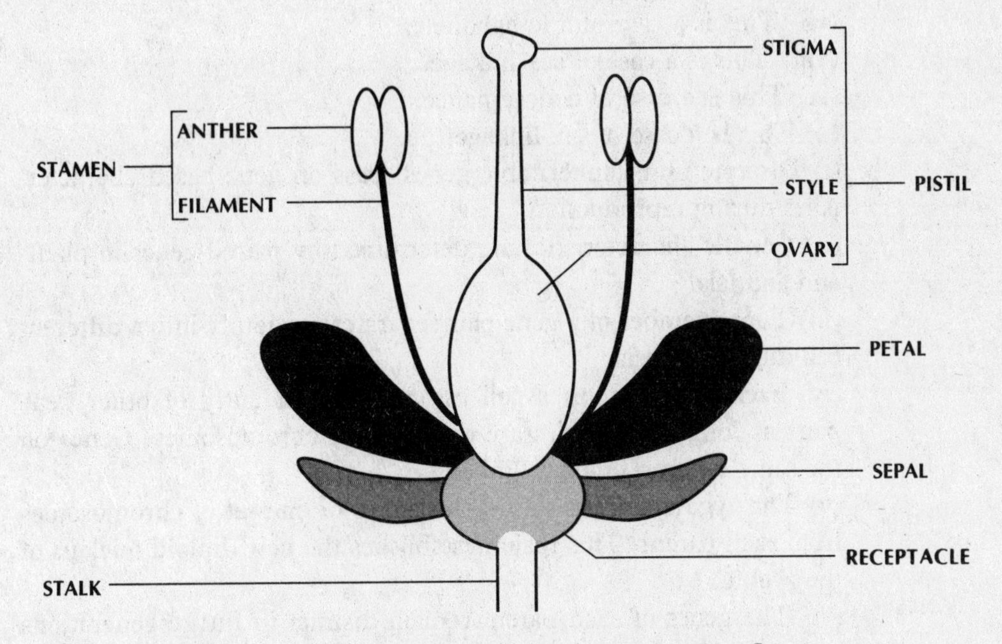

Fig. e. Typical longitudinal section through a flower.

Question 3

a. (i) This is a case of a lethal allele.
(ii) This is a case of multiple loci (polygenes).
(iii) This is a case of nondisjunction.
(iv) This is a case of multiple alleles.
(v) This is a case of sex-linkage.
(vi) This is a case of a recessive allele.
(vii) This is a case of a lethal allele.
(viii) This is a case of sex-linkage.
(ix) This is a case of codominance.
(x) This is a case of sex-linkage.

b. (i) Discrete units, inheritable genes, pass on gene-based characteristics during reproduction.
(ii) Genetic characteristics are determined by paired genes in plants and animals.
(iii) Each member of a gene pair separates randomly into a different cell during meiosis.
(iv) Each gene pair in a cell assorts independently of other gene pairs as long as the pairs are on different chromosomes. Genes on the same chromosome are linked.
(v) The zygote is the result of the fusion of one set of chromosomes from each parent. This fusion establishes the new diploid nucleus of the zygote.
(vi) The genes of each parent remain distinct in future generations and may either continue or appear and disappear again and again in succeeding generations.

Question 4

a. Chloroplasts are surrounded by a double membrane, one outer and the other inner. The inside of the chloroplast is composed of another set of membranes that form stacked units called thylokoids. Thylokoids are connected to one another by means of membranous lamellae. The thylokoids and lamallae are surrounded by a ground substance material called stroma. The thylokoid membranes contain the chlorophyll molecules and enzymes necessary for photosystems I and II to make energy and hydrogen available for the dark reaction. The enzymes for the dark reaction are located in the stroma.

Mitochondria are also bounded by a double membrane. The outer membrane surrounds the mitochondrion while the inner membrane is thrown into internal folds called cristae. The inner portion of the mitochondrion is filled with a ground substance that forms a matrix. Some of the enzymes for the tricarboxylic acid cycle are located in this matrix while the other enzymes are bound to the cristae membranes. The cristae membranes also contain the enzymes necessary for the respiratory chain chemistry.

b. Both mitochondria and plastids contain their own DNA, RNA, and ribosomes. Mitochondria and plastids are capable of self-reproduction. Cells are apparently incapable of making mitochondria or plastids from raw materials. These facts support the hypothesis that early in evolution, mitochondria and plastids were prokaryotes that found it advantageous to take up residence in eukaryotic cells.

c. (i) Plant cells have cell walls while animal cells do not.
(ii) Plant cells possess plastids while animal cells do not.
(iii) Plant cells possess tonoplasts while animal cells do not.

The examination that follows offers a third chance for you to assess your readiness for the Advanced Placement (AP) Examination in Biology. Allow yourself 90 minutes to answer all of the 120 multiple-choice questions in Section 1 and 90 minutes to answer the four essays in questions in Section 2. When you have completed the exam, check your answers against the Answer Key and Explanatory Answers at the end of the exam. Complete the Diagnostic Chart on page 247 to identify any remaining weaknesses.

Use the specially constructed answer sheet to record your answers for Section 1. Use plain or lined paper to answer the free-response questions in Section 2. Sample answers are provided for each essay question.

MODEL EXAMINATION 3, SECTION 1
ANSWER SHEET

1 Ⓐ Ⓑ Ⓒ Ⓓ Ⓔ
2 Ⓐ Ⓑ Ⓒ Ⓓ Ⓔ
3 Ⓐ Ⓑ Ⓒ Ⓓ Ⓔ
4 Ⓐ Ⓑ Ⓒ Ⓓ Ⓔ
5 Ⓐ Ⓑ Ⓒ Ⓓ Ⓔ
6 Ⓐ Ⓑ Ⓒ Ⓓ Ⓔ
7 Ⓐ Ⓑ Ⓒ Ⓓ Ⓔ
8 Ⓐ Ⓑ Ⓒ Ⓓ Ⓔ
9 Ⓐ Ⓑ Ⓒ Ⓓ Ⓔ
10 Ⓐ Ⓑ Ⓒ Ⓓ Ⓔ
11 Ⓐ Ⓑ Ⓒ Ⓓ Ⓔ
12 Ⓐ Ⓑ Ⓒ Ⓓ Ⓔ
13 Ⓐ Ⓑ Ⓒ Ⓓ Ⓔ
14 Ⓐ Ⓑ Ⓒ Ⓓ Ⓔ
15 Ⓐ Ⓑ Ⓒ Ⓓ Ⓔ
16 Ⓐ Ⓑ Ⓒ Ⓓ Ⓔ
17 Ⓐ Ⓑ Ⓒ Ⓓ Ⓔ
18 Ⓐ Ⓑ Ⓒ Ⓓ Ⓔ
19 Ⓐ Ⓑ Ⓒ Ⓓ Ⓔ
20 Ⓐ Ⓑ Ⓒ Ⓓ Ⓔ
21 Ⓐ Ⓑ Ⓒ Ⓓ Ⓔ
22 Ⓐ Ⓑ Ⓒ Ⓓ Ⓔ
23 Ⓐ Ⓑ Ⓒ Ⓓ Ⓔ
24 Ⓐ Ⓑ Ⓒ Ⓓ Ⓔ
25 Ⓐ Ⓑ Ⓒ Ⓓ Ⓔ
26 Ⓐ Ⓑ Ⓒ Ⓓ Ⓔ
27 Ⓐ Ⓑ Ⓒ Ⓓ Ⓔ
28 Ⓐ Ⓑ Ⓒ Ⓓ Ⓔ
29 Ⓐ Ⓑ Ⓒ Ⓓ Ⓔ
30 Ⓐ Ⓑ Ⓒ Ⓓ Ⓔ
31 Ⓐ Ⓑ Ⓒ Ⓓ Ⓔ
32 Ⓐ Ⓑ Ⓒ Ⓓ Ⓔ
33 Ⓐ Ⓑ Ⓒ Ⓓ Ⓔ
34 Ⓐ Ⓑ Ⓒ Ⓓ Ⓔ
35 Ⓐ Ⓑ Ⓒ Ⓓ Ⓔ
36 Ⓐ Ⓑ Ⓒ Ⓓ Ⓔ
37 Ⓐ Ⓑ Ⓒ Ⓓ Ⓔ
38 Ⓐ Ⓑ Ⓒ Ⓓ Ⓔ
39 Ⓐ Ⓑ Ⓒ Ⓓ Ⓔ
40 Ⓐ Ⓑ Ⓒ Ⓓ Ⓔ
41 Ⓐ Ⓑ Ⓒ Ⓓ Ⓔ
42 Ⓐ Ⓑ Ⓒ Ⓓ Ⓔ
43 Ⓐ Ⓑ Ⓒ Ⓓ Ⓔ
44 Ⓐ Ⓑ Ⓒ Ⓓ Ⓔ
45 Ⓐ Ⓑ Ⓒ Ⓓ Ⓔ
46 Ⓐ Ⓑ Ⓒ Ⓓ Ⓔ
47 Ⓐ Ⓑ Ⓒ Ⓓ Ⓔ
48 Ⓐ Ⓑ Ⓒ Ⓓ Ⓔ
49 Ⓐ Ⓑ Ⓒ Ⓓ Ⓔ
50 Ⓐ Ⓑ Ⓒ Ⓓ Ⓔ
51 Ⓐ Ⓑ Ⓒ Ⓓ Ⓔ
52 Ⓐ Ⓑ Ⓒ Ⓓ Ⓔ
53 Ⓐ Ⓑ Ⓒ Ⓓ Ⓔ
54 Ⓐ Ⓑ Ⓒ Ⓓ Ⓔ
55 Ⓐ Ⓑ Ⓒ Ⓓ Ⓔ
56 Ⓐ Ⓑ Ⓒ Ⓓ Ⓔ
57 Ⓐ Ⓑ Ⓒ Ⓓ Ⓔ
58 Ⓐ Ⓑ Ⓒ Ⓓ Ⓔ
59 Ⓐ Ⓑ Ⓒ Ⓓ Ⓔ
60 Ⓐ Ⓑ Ⓒ Ⓓ Ⓔ
61 Ⓐ Ⓑ Ⓒ Ⓓ Ⓔ
62 Ⓐ Ⓑ Ⓒ Ⓓ Ⓔ
63 Ⓐ Ⓑ Ⓒ Ⓓ Ⓔ
64 Ⓐ Ⓑ Ⓒ Ⓓ Ⓔ
65 Ⓐ Ⓑ Ⓒ Ⓓ Ⓔ
66 Ⓐ Ⓑ Ⓒ Ⓓ Ⓔ
67 Ⓐ Ⓑ Ⓒ Ⓓ Ⓔ
68 Ⓐ Ⓑ Ⓒ Ⓓ Ⓔ
69 Ⓐ Ⓑ Ⓒ Ⓓ Ⓔ
70 Ⓐ Ⓑ Ⓒ Ⓓ Ⓔ
71 Ⓐ Ⓑ Ⓒ Ⓓ Ⓔ
72 Ⓐ Ⓑ Ⓒ Ⓓ Ⓔ
73 Ⓐ Ⓑ Ⓒ Ⓓ Ⓔ
74 Ⓐ Ⓑ Ⓒ Ⓓ Ⓔ
75 Ⓐ Ⓑ Ⓒ Ⓓ Ⓔ
76 Ⓐ Ⓑ Ⓒ Ⓓ Ⓔ
77 Ⓐ Ⓑ Ⓒ Ⓓ Ⓔ
78 Ⓐ Ⓑ Ⓒ Ⓓ Ⓔ
79 Ⓐ Ⓑ Ⓒ Ⓓ Ⓔ
80 Ⓐ Ⓑ Ⓒ Ⓓ Ⓔ
81 Ⓐ Ⓑ Ⓒ Ⓓ Ⓔ
82 Ⓐ Ⓑ Ⓒ Ⓓ Ⓔ
83 Ⓐ Ⓑ Ⓒ Ⓓ Ⓔ
84 Ⓐ Ⓑ Ⓒ Ⓓ Ⓔ
85 Ⓐ Ⓑ Ⓒ Ⓓ Ⓔ
86 Ⓐ Ⓑ Ⓒ Ⓓ Ⓔ
87 Ⓐ Ⓑ Ⓒ Ⓓ Ⓔ
88 Ⓐ Ⓑ Ⓒ Ⓓ Ⓔ
89 Ⓐ Ⓑ Ⓒ Ⓓ Ⓔ
90 Ⓐ Ⓑ Ⓒ Ⓓ Ⓔ
91 Ⓐ Ⓑ Ⓒ Ⓓ Ⓔ
92 Ⓐ Ⓑ Ⓒ Ⓓ Ⓔ
93 Ⓐ Ⓑ Ⓒ Ⓓ Ⓔ
94 Ⓐ Ⓑ Ⓒ Ⓓ Ⓔ
95 Ⓐ Ⓑ Ⓒ Ⓓ Ⓔ
96 Ⓐ Ⓑ Ⓒ Ⓓ Ⓔ
97 Ⓐ Ⓑ Ⓒ Ⓓ Ⓔ
98 Ⓐ Ⓑ Ⓒ Ⓓ Ⓔ
99 Ⓐ Ⓑ Ⓒ Ⓓ Ⓔ
100 Ⓐ Ⓑ Ⓒ Ⓓ Ⓔ
101 Ⓐ Ⓑ Ⓒ Ⓓ Ⓔ
102 Ⓐ Ⓑ Ⓒ Ⓓ Ⓔ
103 Ⓐ Ⓑ Ⓒ Ⓓ Ⓔ
104 Ⓐ Ⓑ Ⓒ Ⓓ Ⓔ
105 Ⓐ Ⓑ Ⓒ Ⓓ Ⓔ
106 Ⓐ Ⓑ Ⓒ Ⓓ Ⓔ
107 Ⓐ Ⓑ Ⓒ Ⓓ Ⓔ
108 Ⓐ Ⓑ Ⓒ Ⓓ Ⓔ
109 Ⓐ Ⓑ Ⓒ Ⓓ Ⓔ
110 Ⓐ Ⓑ Ⓒ Ⓓ Ⓔ
111 Ⓐ Ⓑ Ⓒ Ⓓ Ⓔ
112 Ⓐ Ⓑ Ⓒ Ⓓ Ⓔ
113 Ⓐ Ⓑ Ⓒ Ⓓ Ⓔ
114 Ⓐ Ⓑ Ⓒ Ⓓ Ⓔ
115 Ⓐ Ⓑ Ⓒ Ⓓ Ⓔ
116 Ⓐ Ⓑ Ⓒ Ⓓ Ⓔ
117 Ⓐ Ⓑ Ⓒ Ⓓ Ⓔ
118 Ⓐ Ⓑ Ⓒ Ⓓ Ⓔ
119 Ⓐ Ⓑ Ⓒ Ⓓ Ⓔ
120 Ⓐ Ⓑ Ⓒ Ⓓ Ⓔ

MODEL EXAMINATION 3

SECTION 1

Time—90 minutes

Number of Questions—120

Percent of Total Grade—40

Directions: **For each of the 120 questions or incomplete statements below, select the choice from (A–E) that best answers the question or completes the statement. Record your answers on the answer sheet provided.**

1. A poison that destroys a particular enzyme in the Krebs Citric Acid Cycle would most likely affect cells by
   (A) killing them immediately
   (B) causing them to enter fermentation for a short period of time and then causing their death
   (C) causing them to enter fermentation for an extended period of time followed by a period of recovery
   (D) slowing down the synthesis of enzymes important in the production of carbohydrates
   (E) causing them to immediately enter mitosis so as to dilute the effect of the poison

2. The structure that is common to both the digestive and respiratory systems is the
   (A) larynx
   (B) pharynx
   (C) esophagus
   (D) stomach
   (E) trachea

3. The term chemical evolution refers to the
   (A) way in which heterotrophic cells evolved into autotrophic cells
   (B) sequence of events in which atoms evolved into complex aggregates such as lipids
   (C) sequence of events in the evolution of the vertebrates
   (D) way in which organic molecules evolved into multicellular organisms
   (E) way in which DNA evolved into genes

4. The interaction between organisms requiring a single common resource is known as
   (A) predation
   (B) competition
   (C) symbiosis
   (D) parasitism
   (E) mutualism

5. In nature, most bacteria and fungi play an important role in
   (A) disease production
   (B) decomposition of dead organisms
   (C) lactic acid and alcohol fermentation
   (D) synthesis of food for producers
   (E) the fixation of nitrogen

6. Common decomposers are the
   (A) mosses
   (B) bryophytes
   (C) angiosperms
   (D) fungi
   (E) chlorophytes

7. One major difference between the aves and mammalia is that the aves
   (A) are poikilotherms
   (B) produce nitrogen waste in the form of uric acid crystals
   (C) exchange gases through alveoli in the lungs
   (D) have a three-chambered heart
   (E) have an open circulatory system

8. Under anaerobic conditions, certain cells can
   (A) convert pyruvic acid into either lactic acid or alcohol
   (B) produce ATP in the presence of oxygen
   (C) convert photosystems I and II into catabolic pathways
   (D) shunt glucose directly into the Krebs Cycle
   (E) show a decrease in the number of mitochondria per unit volume of cell

9. Movements that are light dependent are called
   (A) chemotaxes
   (B) geotaxes
   (C) phototaxes
   (D) chronotaxes
   (E) thermotaxes

10. The study of the phenomenon of imprinting is associated with the name
    (A) Charles Darwin
    (B) Sidney Fox
    (C) Benjamin Skinner
    (D) Robert Koch
    (E) none of these

11. Acid chyme stimulates the duodenal wall to secrete the hormones
    (A) pancreatic amylase and pancreatic trypsin
    (B) pepsin and rennin
    (C) secretin and choleocystokinin
    (D) cortisone and hydrocortisone
    (E) lipase and amylase

12. Gregor Mendel did not study
    (A) multiple alleles
    (B) independent assortment
    (C) dihybrid crosses
    (D) dominant and recessive traits
    (E) monohybrid crosses

13. The best definition of a cell is
    (A) that which contains DNA and RNA
    (B) units that are composed of proteins
    (C) a structure that can grow
    (D) things that are very, very small
    (E) a unit of biological activity, delimited by a semipermeable membrane and capable of self-reproduction in a medium free of other living systems

14. Communication and behavior in bees has been extensively studied by the Nobel Prize winner
    (A) James Watson
    (B) Frederick Sanger
    (C) Karl von Frisch
    (D) Hans Krebs
    (E) Peter Mitchell

15. The body cavity that is lined by tissues derived from the mesoderm is referred to as the
    (A) archenteron
    (B) coelom
    (C) blastocoel
    (D) central canal
    (E) blastocyst cavity

16. Animal behavior is often not alterable by
    (A) chemical communication
    (B) visual communication
    (C) auditory communication
    (D) tactile communication
    (E) none of these

17. The ploidy statement that is untrue for angiosperms is
    (A) monocot and dicot mesophyll cells are generally all diploid

(B) megaspores are generally diploid
(C) microspores are haploid
(D) endosperm cells are triploid
(E) cortex cells in monocots and dicots are generally diploid

18. A gland cell capable of producing large quantities of protein hormone would have well-developed
(A) cilia
(B) centrioles
(C) rough endoplasmic reticulum
(D) smooth endoplasmic reticulum
(E) chromosomes

19. Normally, the cytoplasm of a cell does not contain
(A) chloroplasts
(B) mitochondria
(C) ribosomes
(D) lysosomes
(E) chromosomes

20. Given the following information, AA = CURLY HAIR, AA′ = WAVY HAIR, and A′A′ = STRAIGHT HAIR, one would suspect that the relationship between A and A′ is that of
(A) codominance
(B) incomplete dominance
(C) sex-linkage
(D) dominant–recessive
(E) multiple of alleles

21. The three types of natural selection operating within natural populations are
(A) stabilizing selection, disruptive selection, and directional selection
(B) stabilizing selection, diminished selection, and directional selection
(C) disruptive selection, diminished selection, and stabilizing selection
(D) qualitative selection, quantitative selection, and compressive selection
(E) disruptive selection, directional selection, and compressive selection

22. The famous Austrian ethologist who first "imprinted" ducks and goslings to himself is
(A) Gregor Mendel
(B) Charles Darwin
(C) Carl von Frisch
(D) Thomas Morgan
(E) Konrad Lorenz

23. Genetic "shuffling" (crossing over) occurs during
(A) interphase
(B) synapsis
(C) synapse
(D) metaphase II
(E) anaphase I

24. In higher animals it appears that sexual behavior is generally dependent on
(A) shifts in the level of hormones in response to other factors
(B) rises and declines in body temperature due to paralysis of the hypothalamus
(C) changes in blood pH due to changes in the length of daylight
(D) the number of males in a given area
(E) changes in the animal's diet

25. Male gametophytes in flowering plants develop in the
(A) anthers
(B) stylus
(C) ovaries
(D) filaments
(E) petals

26. The estrous cycle is the result of a complex feedback mechanism involving the
(A) anterior pituitary, posterior pituitary, and the ovary
(B) posterior pituitary, ovary, and the hypothalamus
(C) uterus, ovary, and the anterior pituitary
(D) anterior pituitary, ovary, and the hypothalamus
(E) uterus, ovary, and the hypothalamus

27. An external skeleton composed of chitin and the presence of jointed appendages to aid in locomotion is characteristic of Phylum
    (A) Platyhelminthes
    (B) Annelida
    (C) Chordata
    (D) Echinodermata
    (E) Arthropoda

28. The canal that allows for pressure equalization in the middle ear is the
    (A) semicircular canal
    (B) eustachian tube
    (C) external auditory meatus
    (D) external nares
    (E) malleus

29. The best explanation of why linked genes do not obey Mendel's Law of Independent Assortment is that
    (A) linked alleles are on the same chromosome
    (B) linked alleles mutate much too rapidly to keep track of
    (C) linked alleles are usually lethal
    (D) Mendel's Law of Independent Assortment only applies to pea plants
    (E) the basic structure of DNA in plants is very different from the basic structure of DNA in animals

30. A hemophiliac married to a homozygous normal woman can expect what percentage of his daughters will be carriers for hemophilia?
    (A) 0%
    (B) 25%
    (C) 50%
    (D) 75%
    (E) 100%

31. Which of the following habitats contains the greatest number of xerophytes?
    (A) oceans
    (B) lakes
    (C) deserts
    (D) deciduous forests
    (E) tropical rain forests

32. Flukes, tapeworms, and planaria are typical organisms in the Phylum
    (A) Platyhelminthes
    (B) Annelida
    (C) Chordata
    (D) Echinodermata
    (E) Mollusca

Questions 33 and 34. The anticancer drug Bleomycin and the free amino acid Taurine were used in a study of a strain of mice that has a dominant gene for the production of tumors. The result of the research is presented in the bar graph below.

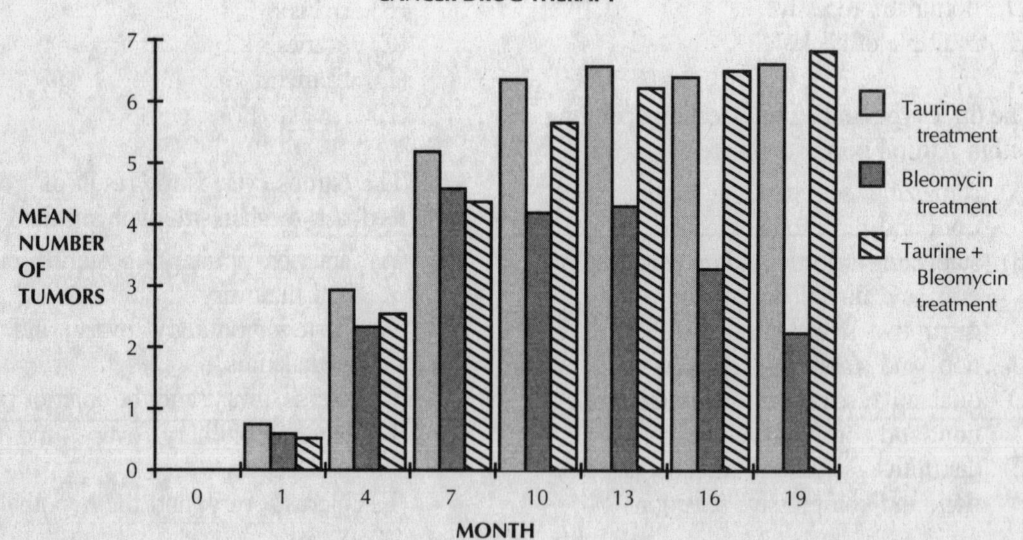

33. From the data it is clear that mice treated with Bleomycin only had fewest number of tumors in week
    (A) 1 (B) 19 (C) 10 (D) 14 (E)6

34. Which of the following conclusions is the most reasonable from the data presented in the graph?
    (A) Bleomycin is effective in reducing the number of tumors.
    (B) There is a strong correlation between body temperature and the production of tumors in the mice.
    (C) Tumors in bone and cartilage appear to be uninfluenced by Bleomycin.
    (D) Bleomycin and Taurine work together in reducing the number of tumors that develop.
    (E) Taurine has potential as an anticancer drug.

35. The part of the flowering plant that is referred to as wood is
    (A) primary xylem
    (B) secondary xylem
    (C) primary phloem
    (D) secondary phloem
    (E) pith

36. Synaptic vesicles are
    (A) released at the end of dendrites and axons
    (B) released at the centriole end of the nucleus
    (C) released at the end of axons only
    (D) produced in the nucleus
    (E) synthesized in the thylakoid membranes

37. The lowest possible blood pressure would be found in a/an
    (A) artery (B) arteriole (C) capillary (D) venule (E) vein

38. When a compound is oxidized, it
    (A) loses electrons
    (B) gains electrons
    (C) loses protons
    (D) gains protons
    (E) all of the above

39. The pollen grain is the source of the
    (A) sperm nucleus
    (B) egg nucleus
    (C) florigen
    (D) stimulus for stamen lengthening
    (E) prolactin

40. The statement(s) that is (are) true of the breathing center in mammals is (are)
    (A) The breathing center is located in the medulla oblongata.
    (B) The breathing center sends motor impulses to the muscles of the ribs and diaphragm.
    (C) The breathing center contains sensors for carbon dioxide concentration in the blood.
    (D) all of the above
    (E) (A and B) only

41. A base, when added to water, will
    (A) lower the pH of the water
    (B) neutralize all of the water
    (C) raise the pH of the water
    (D) acidify the water
    (E) dehydrate the water

42. When one of the pollen nuclei combines with the two polar nuclei of the ovule, the resultant cell develops into the
    (A) endosperm (B) radicle (C) embryo (D) plummule (E) cortex

43. Forms of the same element that differ in the number of neutrons present are
    (A) ions (B) cations (C) isomers (D) isotopes (E) anions

44. In the period during which anaphase II of meiosis is occurring
    (A) tetrads form and then separate
    (B) chromatin threads coil up into chromosomes

(C) chromatids separate
(D) centrioles move to the poles
(E) spindle fibers begin to be assembled

45. One function that is not attributed to the kidney is
(A) excretion of nitrogenous wastes
(B) regulation of heart rate
(C) regulation of blood pH
(D) regulation of water volume in the blood
(E) maintenance of ion concentration in the blood

46. In the angiospermae, sexual reproduction involves
(A) roots and stems
(B) pollen and ovules
(C) rhizomes and rhizoids
(D) monocots and dicots
(E) sepals and petals

Questions 47–49. The graph below is an absorption spectrum determined with chlorophyll extracted from spinach leaves.

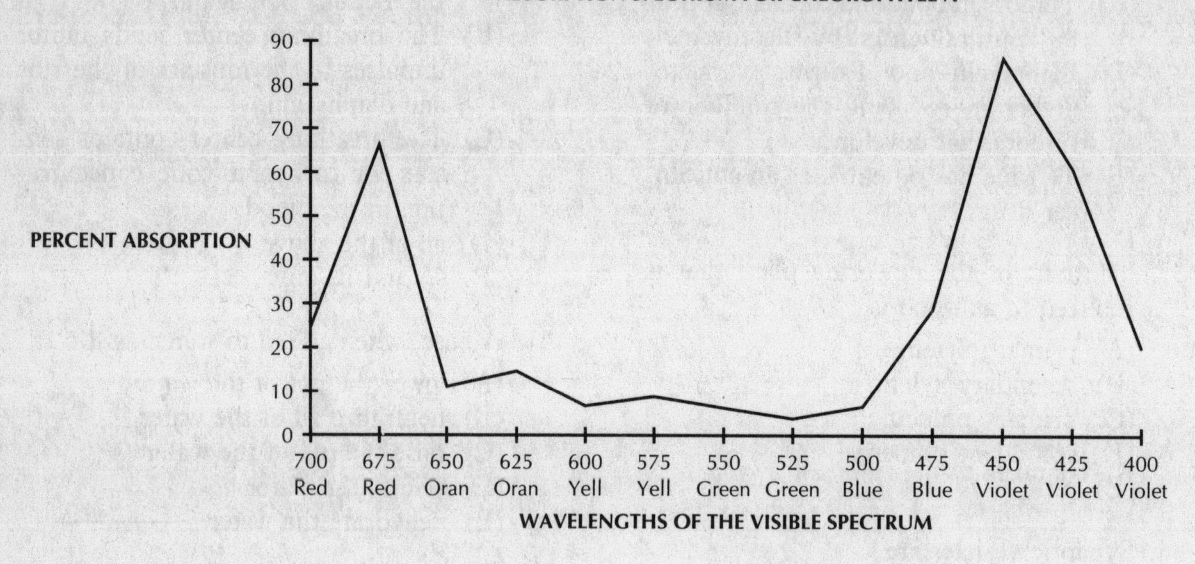

47. The wavelength of visible light that would least contribute energy to the process of photosynthesis is
(A) red (B) yellow (C) green (D) blue (E) violet

48. The wavelength of visible light that would most contribute energy to the process of photosynthesis is
(A) red (B) yellow (C) green (D) blue (E) violet

49. Which statement does the data in the graph support?
(A) Red and violet light are best absorbed by chlorophyll.
(B) Red and violet light are least absorbed by chlorophyll.
(C) Green light alone is the best absorbed by chlorophyll.
(D) Chlorophyll receives its energy from ultraviolet light.
(E) Chlorophyll absorbs all wavelengths of visible light equally.

50. The chemical properties of an atom are essentially due to the number of
(A) electrons in the outer shell
(B) electrons in the innermost shell
(C) protons in its outer shell
(D) neutrons in its outer shell
(E) neutrons in the atomic nucleus

51. With regard to relative size, arrange (smallest to large) the following:
    (1) electron
    (2) nucleic acid molecule
    (3) carbon atom
    (4) fructose molecule
    (5) water molecule

    (A) (2)–(4)–(5)–(3)–(1)
    (B) (4)–(2)–(1)–(3)–(5)
    (C) (4)–(1)–(3)–(2)–(5)
    (D) (1)–(3)–(5)–(4)–(2)
    (E) (2)–(4)–(3)–(5)–(1)

52. Which statement is *not* true?
    (A) Normally, photosynthesis cannot occur at night.
    (B) Photosynthesis requires a supply of carbon dioxide.
    (C) Photosynthesis can be considered an anabolic form of metabolism.
    (D) Photosynthesis represents an oxidation reaction.
    (E) Photosynthetic rate is influenced by temperature.

53. The last portion of the small intestine
    (A) produces pepsin
    (B) connects to the stomach
    (C) absorbs digested foods
    (D) primarily absorbs water and minerals
    (E) both (A and B)

54. Plant cell mitosis differs from animal cell mitosis in that
    (A) plant cell division occurs without aster formation
    (B) plant cell division occurs without spindle formation
    (C) diploid animal cells produce haploid spores
    (D) haploid plant cells produce diploid spores
    (E) plant cells develop a cleavage furrow

55. During spermatogenesis
    (A) mitosis occurs once
    (B) meiosis occurs once
    (C) mitosis occurs twice
    (D) meiosis occurs twice
    (E) mitosis and meiosis alternate twice

56. During secondary growth in dicot stems, the fate of secondary phloem is to
    (A) become part of the pith
    (B) be converted into the pericycle
    (C) be crushed against the inside of the bark
    (D) become sapwood
    (E) be converted into secondary xylem

57. Phytochrome is a
    (A) cofactor required for chlorophyll functioning
    (B) specific enzyme found in both photosystems I and II
    (C) pigment required for flowering
    (D) cofactor in the Calvin–Benson Cycle
    (E) common pigment found in starch granules

58. It is believed that the primitive atmosphere of earth did not contain
    (A) methane gas
    (B) ammonia
    (C) hydrogen gas
    (D) molecular oxygen
    (E) water vapor

59. The supply of electrons for photosystem II is
    (A) water molecules
    (B) molecular oxygen
    (C) sunlight energy
    (D) the Calvin–Benson Cycle
    (E) photosystem

60. In the human fetus, the vessel carrying the blood with the greatest amount of oxygen is the
    (A) venus duct
    (B) pulmonary artery
    (C) umbilical artery
    (D) umbilical vein
    (E) none of the preceding

Questions 61–63. The bar graph below depicts the distribution of two species of barnacles in the same intertidal zone.

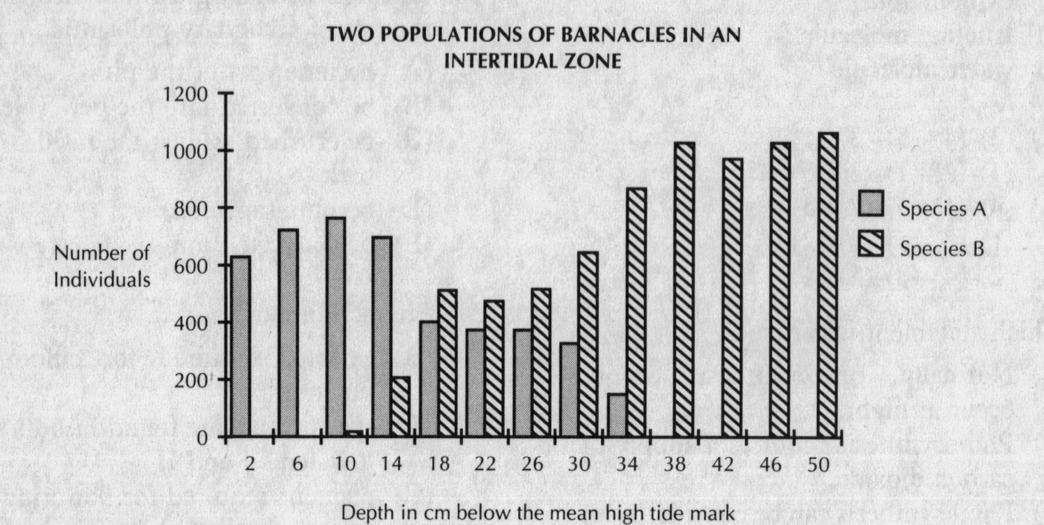

61. The best description of the environmental relationship shown in the graph is
    (A) competition
    (B) mutualism
    (C) parasitism
    (D) imprinting
    (E) commensalism

62. The mean depth at which the two species interact is
    (A) 46 cm (C) 27 cm (E) 6 cm
    (B) 34 cm (D) 24 cm

63. The data reported above support which statement?
    (A) The presence of two species of barnacle precludes the possibility of a third species being present.
    (B) Intertidal zones are the only aquatic environments free of competition.
    (C) Both species of barnacle occupy a common niche.
    (D) Food and temperature influence barnacle distribution.
    (E) Barnacles require sunlight in order to reproduce.

64. In photosystem 11, electrons are transferred to
    (A) ADP (C) cyclic AMP (E) NAD
    (B) P680 (D) NADP

65. According to the cohesion–tension theory of xylem transport
    (A) mass–flow is the key mechanism
    (B) water is pushed up through the stem
    (C) water plays no role in transpiration
    (D) water is pulled up
    (E) both (A and B)

66. A eukaryotic cell differs from a prokaryotic cell in that
    (A) one has a cell membrane while the other has a cell wall
    (B) one has mitochondria with thylakoids while the other does not
    (C) one has observable organelles and the other does not
    (D) one is found in multicellular plants while the other is found in multicellular animals
    (E) both (A and B)

67. The structure primarily found in thylakoid membranes is the
    (A) glycolytic enzyme
    (B) basement membrane
    (C) chlorophyll pigment
    (D) golgi apparatus
    (E) stromal lamella

68. The stomata are structures found in
    (A) the monocot leaf
    (B) xylem of woody plants
    (C) the dicot leaf
    (D) both (A and B)
    (E) both (A and C)

69. A cell in hypertonic solution will
    (A) shrink
    (B) swell
    (C) remain unchanged
    (D) carry on active transport
    (E) none of the above

70. Two important proteolytic enzymes are
    (A) lipase and amylase
    (B) salivary amylase and pancreatic amylase
    (C) estrogen and testosterone
    (D) pepsin and trypsin
    (E) salivary amylase and ptylin

71. The release of pancreatic juice is regulated by
    (A) secretin
    (B) secretin
    (C) amylase
    (D) amylase
    (E) pepsin

72. The mass–flow (pressure–flow) hypothesis of phloem transport incorporates the
    (A) flow of water from the source to a sink
    (B) ATP facilitated transport of sugars
    (C) passive transport in the form of osmosis
    (D) none of the above
    (E) (A, B, and C)

73. A fern is a plant that
    (A) has only xylem but no phloem
    (B) has both xylem and phloem
    (C) lacks vascular tissues
    (D) produces flowers without stigmas
    (E) produces sperm that migrate

Questions 74–77. For these questions, give identifications based on the following food chain.

(A) grass
|
(B) insect
|
(C) mouse
|
(D) snake
|
(E) hawk

74. primary consumer

75. producer

76. tertiary consumer

77. herbivore

78. Bile produced by the liver serves as a/an
    (A) emulsifier of fats
    (B) foundation for red blood cell production in the bone marrow
    (C) cofactor in blood clotting
    (D) enzyme for protein digestion
    (E) substrate for hemoglobin production

GO ON TO THE NEXT PAGE.

Questions 79 and 80 are based on the following chart.

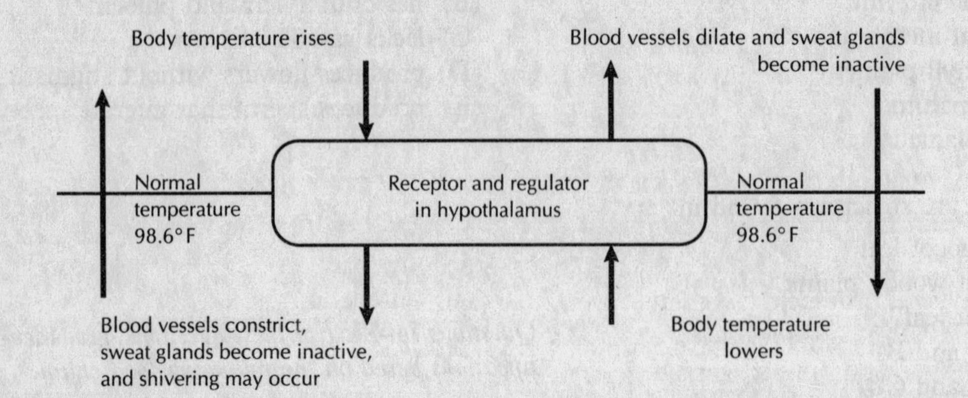

79. The chart best illustrates
    (A) the thermostat in a building
    (B) a type of feedback mechanism
    (C) the phenomenon of homeostasis
    (D) both (B and C)
    (E) none of the above

80. Based on the information in the chart, which statement is correct?
    (A) Arteries and veins are uninfluenced by body temperatures.
    (B) The hypothalamus controls heart rate and respiration.
    (C) The body temperature is under the control of the endocrine system.
    (D) The hypothalamus serves as the center for body temperature regulation.
    (E) All bodily functions are controlled by the hypothalamus.

81. The carbon dioxide present in the primitive atmosphere
    (A) made possible the evolution of heterotrophs
    (B) made possible the evolution of autotrophs
    (C) killed all of the early autotrophs
    (D) made possible the evolution of cellular respiration
    (E) none of the above

82. Depletion curves can be applied to
    (A) nonrenewable resources
    (B) renewable resources
    (C) waste production by humans
    (D) sunlight
    (E) population growth

83. Down's syndrome is the result of
    (A) sex-linked inheritance
    (B) multiple allelic inheritance
    (C) nondisjunction
    (D) simple Mendelian inheritance
    (E) Tay–Sachs disease

84. In human development, the ectoderm gives rise to the
    (A) digestive system
    (B) muscles of the body
    (C) bones of the body
    (D) central nervous system
    (E) excretory system

85. Both steroid and protein hormones affect the rate of
    (A) impulse transmission in motor nerves
    (B) cellular metabolism
    (C) cyclic AMP production
    (D) binding to cell membrane receptor sites
    (E) none of the above

86. One of the anterior pituitary hormones influences the
    (A) kidney
    (B) hypothalamus
    (C) adrenal medulla
    (D) adrenal cortex
    (E) stomach

87. The correct sequence in the life cycle of the moss is
    (A) spore, zygote, protonema, gamete, gametophyte
    (B) gametophyte, sporophyte, gamete, spore, protonema
    (C) zygote, protonema, spore, gamete, gametophyte
    (D) sporophyte, spore, protonema, gamete, zygote
    (E) gametophyte, gamete, sporophyte, spore, protonema

88. The major land plants are
    (A) bryophyta and gymnospermae
    (B) angiospermae and gymnospermae
    (C) rhodophyta and chlorophyta
    (D) bryophyta and chlorophyta
    (E) phaeophyta and angiospermae

89. Deuterostomes and protosomes are distinguished from one another on the basis of
    (A) embryonic development
    (B) digestive tract structure
    (C) color pattern
    (D) circulatory system
    (E) size

90. The ductus deferens
    (A) is found in the ovary
    (B) is found in the testis
    (C) becomes erect during sexual excitation
    (D) leads to the ovarian duct
    (E) degenerates after puberty

91. In the cell cycle, the replication of DNA takes place during
    (A) interphase
    (B) prophase
    (C) metaphase
    (D) anaphase
    (E) telophase

92. Glucose metabolism is affected by
    (A) thyroxin
    (B) adrenalin
    (C) insulin
    (D) none of the preceding
    (E) (A, B, and C)

93. The correct sequence for the development of mammals is
    (A) blastula, neurula, gastrula, morula
    (B) morula, blastula, gastrula, neurula
    (C) gastrula, neurula, morula, blastula
    (D) gastrula, blastula, neurula, morula
    (E) none of the above

94. In the concept of environmental resistance, there is an opposition to
    (A) evolution and succession
    (B) habituation and learned behavior
    (C) competition and behavior
    (D) biotic potential and accelerated growth rate
    (E) none of the above

95. The biome that is in the northern most part of the northern hemisphere is the
    (A) desert
    (B) tropical rain forest
    (C) deciduous rain forest
    (D) taiga
    (E) tundra

96. The ozone shield protects the earth from
    (A) infrared radiation
    (B) intense visible light radiation
    (C) ultraviolet light radiation
    (D) harmful greenlight radiation
    (E) invasion by aliens from space

Questions 97 and 98. The graph below reflects the composition of three typical meals.

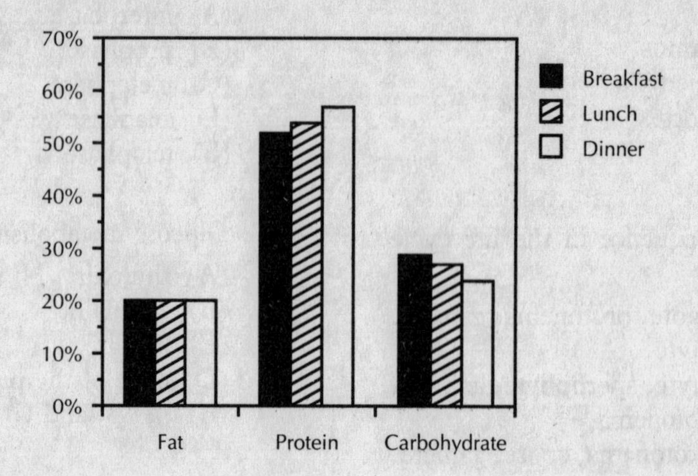

97. Based on the graph, which statement is true?
   (A) The percentage of protein taken in at lunch is lower than the precentage of protein taken in for breakfast.
   (B) The percentage of fat taken in is the same for all three meals.
   (C) The percentage of protein taken in is the same for all three meals.
   (D) The percentage of fat taken in for dinner is greater than the percentage of carbohydrate taken in for breakfast.
   (E) all of the above

98. The meal that is composed of approximately 52% protein is
   (A) breakfast
   (B) lunch
   (C) dinner
   (D) none of the above
   (E) (A and C)

99. The following plant structure that is the true fruit is the
   (A) carrot
   (B) parsley
   (C) cucumber
   (D) onion
   (E) sweet potato

100. The four-chambered heart is typical of
   (A) reptiles
   (B) birds
   (C) amphibians
   (D) arthropods
   (E) fishes

101. The bones that are not part of the axial skeleton are the
   (A) ribs
   (B) vertebrae
   (C) ear ossicles
   (D) facial bones
   (E) carpals

102. Echinoderms are characterized by the presence of a
   (A) closed circulatory system
   (B) tracheal system
   (C) simple gastrovascular cavity
   (D) water vascular system
   (E) heart–lung respiratory system

103. The dominant generation in seed plants is the
   (A) gametophyte
   (B) sporophyte
   (C) xerophyte
   (D) hydrophyte
   (E) mesophyte

104. Excess urine production suggests
    (A) ACTH inhibition
    (B) ADH inhibition
    (C) ACTH production
    (D) ADH production
    (E) FSH inhibition

105. Select the statement that is correct for human development.
    (A) The central nervous system develops much earlier than does the alimentary canal.
    (B) Digits develop prior to the limbs.
    (C) Sense organs develop prior to the nervous system.
    (D) The brain develops before the spinal cord.
    (E) Cephalic development and caudal development proceed together.

106. The events that are in proper sequences are
    (A) mutation—translation—transcription
    (B) mutation—translation
    (C) translation—mutation—transcription
    (D) transcription—mutation
    (E) transcription—translation

107. Increasing human population size naturally leads to
    (A) a reduction in areas of natural resources
    (B) an increase in areas of pollution
    (C) a reduction in habital space and available drinking water
    (D) all of the above
    (E) (A and B) only

108. The first cell was most likely
    (A) a heterotroph
    (B) an autotroph
    (C) a fermenter
    (D) only (A and C)
    (E) only (B and C)

Questions 109–112. These four questions are based on the diagram of a typical sequential biochemical reaction and the letter key, shown below.

```
    1     2     3     4     5
G ___ H ___ J ___ K ___ L ___ M
```

(A) A lipid compound
(B) A specified enzyme
(C) An intermediate compound
(D) The substrate for the reaction
(E) The product of reaction

109. Represented by the number 2

110. Represented by the letter G

111. Represented by the letter J

112. Represented by the letter M

GO ON TO THE NEXT PAGE.

Questions 113 and 114 are based on the following chart.

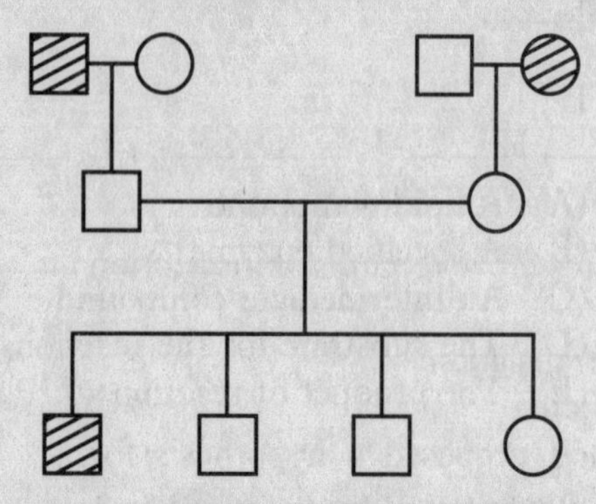

The shaded individuals are afflicted with a simple Mendelian recessive disorder

113. The parents of the four children are clearly
(A) AA × AA
(B) aa × aa
(C) Ae × Ae
(D) Aa × aa
(E) Ae × AA

114. If both parents of the four children have blood type A, they could never have a child with blood type
(A) A
(B) O
(C) B
(D) A or O
(E) both answers (A and C)

115. The vessel that distributes blood to all of the systemic circulation in mammals is the
(A) aorta
(B) common carotid artery
(C) superior vena cava
(D) inferior vena cava
(E) coronary artery

Questions 116–120. These questions are based on the following phyla characteristics key:

(A) Clear segmentation, well defined coelom, and a closed circulation
(B) A tube-within-a-tube body plan and a pseudo coelom
(C) Two germ layers, a saclike body plan, and radial symmetry
(D) A hollow, dorsal nerve cord, gill pouches, and a notochord at some time during the life cycle.
(E) Three germ layers, internal organs, saclike body plan, and bilateral symmetry.

116. Characteristic of Phylum Chordata.

117. Characteristic of the Phylum Annelida.

118. Characteristic of Phylum Cnidaria.

119. Characteristic of the Phylum Nematoda.

120. Characteristic of the Platyhelminthes.

SECTION 2

Time—90 minutes
Percent of Total Grade—40

The questions in this section are mandatory. Answer *ALL* four questions.

1. Protein synthesis is a complex biochemical process that requires transcription and translation.
   a. Explain what is meant by transcription and translation.
   b. Briefly describe how protein synthesis is initiated.
   c. In 1961, Francois Jacob and Jacques Monod proposed a hypothesis for control of protein synthesis in prokaryotes. Summarize their hypothesis.

2. A species is an interbreeding group of organisms that are reproductively isolated from other groups of that species. Reproductive isolation protects a coadapted gene pool that gives rise to a limited number of phenotypes.
   a. define and discuss briefly each of the mechanistic terms listed below.
      (i) allopatric speciation
      (ii) sympatric speciation
      (iii) parapatric speciation
   b. Define the term *prezygotic isolating mechanisms*.
   c. List and describe four main categories of prezygotic species-isolating mechanisms.

3. Describe the structure of the human respiratory system. In addition, explain the mechanism employed in inspiration and expiration.

4. In 1941, Daveson and Danielli proposed a trilaminar structure for the cell membrane. The model consisted of a layer of lipid with a layer of protein on either side.
   a. Describe the current model for the cell membrane.
   b. Describe how things move across the cell membrane.

ANSWER KEY

SECTION I

1. **B**
2. **B**
3. **B**
4. **B**
5. **B**
6. **D**
7. **B**
8. **A**
9. **C**
10. **E**
11. **C**
12. **A**
13. **E**
14. **C**
15. **B**
16. **E**
17. **B**
18. **C**
19. **E**
20. **B**
21. **A**
22. **E**
23. **B**
24. **A**
25. **A**
26. **D**
27. **E**
28. **B**
29. **A**
30. **E**
31. **C**
32. **A**
33. **A**
34. **A**
35. **B**
36. **C**
37. **E**
38. **A**
39. **A**
40. **D**
41. **C**
42. **A**
43. **D**
44. **C**
45. **B**
46. **B**
47. **C**
48. **E**
49. **A**
50. **A**
51. **D**
52. **D**
53. **C**
54. **A**
55. **B**
56. **C**
57. **C**
58. **D**
59. **A**
60. **D**
61. **A**
62. **D**
63. **C**
64. **D**
65. **D**
66. **C**
67. **C**
68. **E**
69. **A**
70. **D**
71. **B**
72. **E**
73. **B**
74. **B**
75. **A**
76. **D**
77. **B**
78. **A**
79. **C**
80. **D**
81. **B**
82. **A**
83. **C**
84. **D**
85. **B**
86. **D**
87. **D**
88. **B**
89. **A**
90. **B**
91. **A**
92. **C**
93. **B**
94. **D**
95. **E**
96. **C**
97. **B**
98. **A**
99. **C**
100. **B**
101. **E**
102. **D**
103. **B**
104. **B**
105. **E**
106. **E**
107. **D**
108. **D**
109. **B**
110. **D**
111. **C**
112. **E**
113. **C**
114. **C**
115. **A**
116. **D**
117. **A**
118. **C**
119. **B**
120. **E**

Diagnostic Chart

Section 1

This chart will provide you with the opportunity of identifying those areas of biology in which you have done well and those areas in biology in which you need to improve your knowledge and understanding. For each correct answer, place an X in the space provided. You are doing well if you have 80–90% of the answers correct in each topic area. You are also doing well if your *total* score is 85% or better.

Topic	Questions on the Examination	Number Correct
Basic Chemistry	38 43 50 51	
Cell Structure and Function	13 18 19 38 41 66 69	
Enzyme Structure and Function	109 110 111 112	
Cell Division	23 44 54 55 91	
Energy Transforms in Living Systems	1 8 52 59 64 67	
Origin of Life	7 27 32 58 81 96	
Taxonomy	88 89 100 102 108	
Higher Plants: Structure and Function	35 39 56 57 65 68 72 87 99	
Higher Plants: Reproduction and Development	17 25 42 46 60 73 103	
Higher Animals: Structure and Function	2 11 28 36 37 40 45 53 70 71 78 85 86 92 101 104 115	
Higher Animals: Reproduction and Development	15 26 84 90 93 105	
Heredity	12 20 29 30 83 106	
Evolution	3 21 116 117 118 119 120	

Topic	Question Numbers	Number Correct
Ecology	4 5 6 31 74 75 76 77 82 94 95 107	______
Behavior	9 10 14 16 22 24	______
Graph, Chart, or Table Interpretation	33 34 47 48 49 61 62 63 79 80 97 98 113 114	______
	Total Correct	______

EXPLANATORY ANSWERS

SECTION 1

1. **(B)** is correct. Fermentation is a short-term alternate pathway for energy production during periods of oxygen deprivation or citric acid cycle shutdown. During extended periods of fermentation there is a buildup of toxic compounds in the cell.

2. **(B)** is correct. The pharynx leads to both the esophagus and the larynx. The entry to the larynx, the glottis, is protected by a flap of tissue known as the epiglottis.

3. **(B)** is correct. Heavy atoms on the cooling primitive earth tended to solidify or liquify. Lighter atoms tended to form gases and vapors. It is these gases and vapors that participated in the formation of simple molecules, simple molecules into complex molecules, and finally complex molecules into super molecules.

4. **(B)** is correct. (A) is the interaction between two species where one species is used as a resource by another species. (C) is the interaction between two species in usually close and long-term association. (D) is a relationship where one species receives benefit at the expense and harm of the other species. (E) is a relationship where both species receive benefit.

5. **(B)** is correct.

6. **(D)** is correct. Fungi are nonphotosynthetic plants that participate in the return of biotic material to the abiotic environment.

7. **(B)** is correct. Mammals generally produce nitrogen wastes in the form of urea. (A) are both homeotherms. (C) both exchange gases through alveoli. (D) both have a four-chambered heart. (E) both have a closed circulatory systems.

8. **(A)** is commonly known as fermentation. (B, C, D, and E) are all nonsense.

9. **(C)** is correct. (A) are movements in response to chemicals. (B) are movements in response to gravitational forces. (D) are movements in response to time. (E) are movements in response to temperature.

10. **(E)** is correct. Konrad Lorenz, the famous ethnologist, studied the phenomenon of imprinting in birds. He is the person who coined the term "imprinting."

11. **(C)** is correct. (A, B, D, and E) are all enzymes, not hormones.

12. **(A)** is correct. Fortunately, Mendel used traits that were simple and therefore allowed him to draw conclusions that would not have been possible if other traits had been selected for his studies.

13. **(E)** is the best definition of a cell.

14. **(C)** is correct. (A) studied the structure of DNA for which he earned a Nobel Prize in 1962. (B) studied the primary structure of insulin and earned a Nobel Prize in 1958. (D) studied the tricarboxylic acid cycle in cellular respiration and earned a Nobel Prize in 1953. (E) studied electron transport mechanisms in the mitochondrial cristae for which he won a Nobel Prize in 1978.

15. **(B)** is correct. (A) is the primitive gut, a cavity lined by endodermal tissues. (C and E) refer to the space within the blastula. (D) is the hollow space within the neural tube.

16. **(E)** is correct. All of the listed forms of communication alter behavior.

17. **(B)** Megaspores are female gametes and are therefore haploid.

18. **(C)** is correct. Protein producing cells would be expected to have well-developed lysosomes for intracellular digestion. One would

also expect a well-developed rough endoplasmic reticulum for protein synthesis, a well-developed golgi apparatus for secretion packaging of hormones, and well-developed mitochondria for all of the cellular energy requirements.

19. **(E)** are organelles that are commonly found in the cell nucleus. (A, B, C, and D) are all cytoplasmic organelles.

20. **(B)** is correct. (A) means that both genes express themselves equally at all times when both are present. (C) has nothing to do with this case since sex chromosomes are not mentioned. (D) would not allow for any expressions other than curly hair or straight hair. (E) are genes that usually exhibit a graded dominant–recessive relationship.

21. **(A)** is correct. (B and C) diminished selection is nonsense. (D) all are nonsense. (E) compressive selection is nonsense.

22. **(E)** is correct. (A) is recognized as the "father" of genetics. (B) is the recognized "father" of evolution. (C) is the recognized "father" of animal behavior studies. (D) postulated the existence of gene linkages.

23. **(B)** is correct. During metaphase I, the paired chromosomes form tetrads. Chromatids exchange genetic materials that provide for new gene combinations.

24. **(A)** is correct. (B, C, D, and E) are nonsense.

25. **(A)** is correct. (B) is the part of the carpel that connects the stigma to the ovary. (C) are the parts of the carpels in which development is initiated in the megaspore mother cell. (D) are the long stalks of the stamens. (E) are the ring of modified leaves within the sepals.

26. **(D)** is correct. Some anterior pituitary hormones trigger changes in the ovary. These changes result in the release of other hormones that feed back to the hypothalamus and trigger changes in releaser factors that control the release of hormones in the anterior pituitary.

27. **(E)** is correct. (A) includes flatworms that are nonsegmented, lack a coelom, and have a sac body plan with one opening. (B) consists of roundworms with well-segmented bodies and a well-developed partitioned coelom. (C) consists of animals with either a dorsal notochord or dorsal hollow nerve cord, and gill pouches or gill slits during some part of the life cycle. (D) consists of animals with radial symmetry and a water-vascular system.

28. **(B)** is correct. (A) is a structure found in the inner ear. (C) is the external ear canal that terminates at the tympanic membrane. (D) is the opening into the nose. (E) is the middle ear bone that is between the tympanic membrane and the incus.

29. **(A)** is correct. (B, C, D, and E) are all untrue.

30. **(E)** is correct. The hemophiliac carries the gene for hemophilia on the x-chromosome. It is the only x-chromosome that the male has to contribute to a daughter.

31. **(C)** is correct. Xerophytes are organisms that are adapted to extremely dry environments.

32. **(A)** is correct. (B) is typified by earthworms, sandworms, and leeches. (C) includes the fishes, frogs, reptiles, birds, and mammals. (D) is typified by starfish, sea cucumbers, and sand dollars. (E) includes the octopus, squid, snail, and clam.

33. **(A)** is correct. (B) is four times greater in number than week 1. (C) is nine times greater in number than week 1. (D and E) are not represented on the graph.

34. **(A)** is correct. (B and C) cannot be determined from the information in the graph. (D and E) are contrary to what the graph depicts.

35. **(B)** is correct. (A) only appears in early growth and is an insignificant part of wood. (C, D, and E) contribute nothing to the formation of wood.

36. **(C)** is correct. (A) is only true of axons. (B and D) are unsupported by the available research. (E) relates to chloroplast structure and function.

37. **(E)** is correct. Blood pressure is highest at the arterial end of circulation and is lowest at the venous end.

38. **(A)** is correct. (B) is a reduction. (C and D) do not occur under normal circumstances.

39. **(A)** is correct. (B) is derived from the ovule. (C) is the flowering hormone. (D) is nonsense. (E) is the hormone for milk production.

40. **(D)** is correct.

41. **(C)** is correct. (A and D) are true of acids. (B) is true only in a specific case, pH 7.0. (E) is nonsense.

42. **(A)** is correct. (B, C, and D) are derived from the zygote. (E) is a tissue found in a young dicot stem or in monocot and dicot roots.

43. **(D)** is correct. (A) are charged atoms or molecules. (B) are positively charged ions. (C) are molecules with the same chemical formulae but different chemical structure. (E) are negatively charged ions.

44. **(C)** is correct. (A) is true of metaphase I. (B) is true of anaphase I. (D) is true of prophase I and II. (E) is true of prophase I and II.

45. **(B)** is correct. (A, C, D, and E) are all functions attributed to the kidney.

46. **(B)** is correct. (A) are not reproductive organs. (C) rhizomes are underground stems and rhizoids are primitive rootlike structures. (D) are types of angiosperms. (E) are nonsexual parts of the flower.

47. **(C)** is correct. It is the color that shows the least absorption and is the energy that is least available for photosynthesis.

48. **(E)** is correct. It shows the greatest percentage of absorption when compared to the other wavelengths of visible light.

49. **(A)** is correct.

50. **(A)** is correct. (B) do not participate in atomic bonding. (C and D) are located in the atomic nucleus.

51. **(D)** is correct.

52. **(D)** is correct. The Calvin–Benson Cycle involves a reduction of carbon dioxide in order to produce glucose.

53. **(C)** is correct. (A) occurs in the stomach. (B) is true of the first portion of the small intestine. (D) is true of the large intestine.

54. **(A)** is correct.

55. **(B)** is correct. Meiosis is a two-division process but is considered to be a single event.

56. **(C)** is correct. Secondary xylem remains each year and forms the typical pattern of rings seen in the stem.

57. **(C)** is correct.

58. **(D)** is correct. The primitive atmosphere of earth is believed to have been a reducing atmosphere, not an oxidizing one as it is today.

59. **(A)** is correct. The hydrogen atoms and electrons released by the photolysis of water are used in the reduction of carbon dioxide in the Calvin–Benson Cycle.

60. **(D)** is correct. This vein carries oxygen picked up from the fetal capillaries adjacent to the maternal capillaries of the placenta.

61. **(A)** is correct. (B) is a relationship in which both species benefit. (C) is a relationship in which one species benefits and the other is harmed. (D) is a factor in behavior. (E) is a relationship in which only one species benefits but the other is unharmed.

62. **(D)** is correct. The graph shows that the two species interact at depths between 14 and 34 cm.

63. **(C)** is correct.

64. **(D)** is correct. The electrons of photosystem I usually recycle. (A) is a receptor of light energy, not electrons. (B) is a chlorophyll pigment found in photosystem II. (C) is not found in photosynthesis. (E) is found in cellular respiration (mitochondria).

65. **(D)** is correct. (A) is used to explain phloem transport. (B) is an explanation that cannot account for tall trees. (C) is contrary to current research findings.

66. **(C)** is correct.

67. **(C)** is correct.

68. **(E)** is correct. Stomata are gas exchange openings formed by two adjacent guard cells in the epidermal layer of the leaf.

69. **(A)** is correct. Cells in hypotonic solution swell by means of osmosis. In isotonic solution they remain unchanged. Osmosis is a passive transport.

70. **(D)** is correct. (A) lipase is an enzyme for lipid digestion while amylase is an enzyme for complex carbohydrate digestion. (B) see (A). (C) are steroid sex hormones. (E) ptylin is another name for salivary amylase.

71. **(B)** is correct. (A) is the hormone that stimulates gastric gland release of additional pepsinogen. (C) is an enzyme for complex carbohydrate digestion. (D) is the stomach enzyme used to digest milk protein. (E) is the stomach enzyme used to digest protein.

72. **(E)** is correct.

73. **(B)** is correct. Ferns belong to the vascular plants.

74. **(B)** is correct. (A) is the producer. (C) is the secondary consumer. (D) is a tertiary consumer. (E) is the terminal or final consumer.

75. **(A)** is correct.

76. **(D)** is correct.

77. **(B)** is correct.

78. **(A)** is correct. Bile, the result of the breakdown of hemoglobin, is composed of salts and pigments. The bile salts emulsify fats facilitating fat absorption into the lacteals.

79. **(C)** is correct.

80. **(D)** is correct.

81. **(B)** is correct. Carbon dioxide is important in the process of photosynthesis.

82. **(A)** is correct. (B, C, and E) can be controlled by humans. (D) is not under human control but is dependable.

83. **(C)** is correct. During mitosis, the movement of chromosomes to the poles is unequal and results in gametes that either lack or have an excess of chromosomes. In Down's syndrome one of the gametes has 22 instead of 23 chromosomes.

84. **(D)** is correct. (A, B, C, and E) arise from mesoderm.

85. **(B)** is correct.

86. **(D)** is correct. ACTH releasing-factor produced by the hypothalamus stimulates the anterior pituitary to release ACTH into the blood stream. The ACTH in turn stimulates the adrenal cortex to release its hormones.

87. **(D)** is correct.

88. **(B)** is correct. These are the most advanced vascular plants.

89. **(A)** is correct. In the protostomes the blastopore becomes the mouth while in the deuterostomes the blastopore becomes the anus.

90. **(B)** is correct. It is the tube that connects the epidiymis to the ejaculatory duct. It is also known as the vas deferens.

91. **(A)** is correct. During the interphase the cell

grows, replicates DNA, and then grows again.

92. **(C)** is correct. Insulin promotes blood glucose uptake by the cells of the body.

93. **(B)** is correct.

94. **(D)** is correct. It is the sum total of environmental factors that limit a regional population increase.

95. **(E)** is correct. It is characterized by a lack of trees, a short growing season, cold temperatures, wet conditions because of limited evaporation, and a permafrost.

96. **(C)** is correct. Ozone is a powerful oxidant that interacts with ultraviolet light energy and thereby serves as an effective screen against this biologically damaging radiation.

97. **(B)** is correct. (A) the percentage of protein shown for breakfast is about 52% while that shown for lunch is about 54%. (C and D) are obviously untrue.

98. **(A)** is correct.

99. **(C)** is correct. The fruit is defined as a ripened ovary or the part of the plant that protects the seeds of the plant.

100. **(B)** is correct. Birds and mammals are the only animals that possess the four-chambered heart (double circulation).

101. **(E)** is correct. The carpals are the bones of the wrist.

102. **(D)** is correct. It is a series of channels that conduct water to the tube feet causing them to expand.

103. **(B)** is correct. In the seed plant the gametophyte generation is relegated to a captive existence in the reproductive organ of the sporophyte.

104. **(B)** is correct. ADH is an antidiuretic. (A) would lead to a reduced release of hormone from the adrenal cortex. (C) would lead to an increased release of hormone from the adrenal cortex. (D) would lead to a decreased production of urine. (E) would lead to a reduced estrogen release from the ovary.

105. **(E)** is correct.

106. **(E)** is correct. Transcription is the process of RNA synthesis by DNA while translation is the process of protein synthesis by RNA.

107. **(D)** is correct. Increased numbers bring a negative impact to many areas of the environment

108. **(D)** is correct. Since the early atmosphere is thought to have been a reducing one, oxidative phosphorylation would have been impossible.

Questions 109–112. The biochemical pathway depicted shows a specific enzyme (numbers) for each conversion. The starting compound (letter G) serves as a substrate for the reaction while the final compound (letter M) is considered the product. The others are considered intermediate compounds.

109. **(B)** is correct.

110. **(D)** is correct.

111. **(C)** is correct.

112. **(E)** is correct.

113. **(C)** is correct. Neither parent shows the recessive trait and yet one of the four children does.

114. **(C)** is correct.

115. **(A)** is correct. (B) distributes blood to the head only. (C) serves as the return of blood to the heart from the head. (D) serves as the return of blood to the heart from the lower regions of the body. (E) distributes blood to the walls of the heart.

116. **(D)** is correct.

117. **(A)** is correct.

118. **(C)** is correct.

119. **(B)** is correct.

120. **(E)** is correct.

Model Essays

Section 2

Because subjectivity enters into the grading of any essay, it is not possible to provide *the* answer to a particular essay question. When you compare your answers with the model answers that follow, keep in mind that there are a number of ways in which they could have been written. Pay attention to the content and the organization of the answer; see whether you have followed a similar style for content and organization.

Question 1

a. Transcription is the process of RNA synthesis on a DNA template. During this process the DNA code for the genetic message is written in a complementary form known as messenger RNA (mRNA). Transcription requires the partial uncoiling of the double helix of DNA and the activity of the enzyme RNA polymerase. RNA polymerase pairs individual RNA nucleotides to their complementary bases on one of the uncoiled DNA strands. When the nucleotides are in position, the enzyme facilitates the polymerization of the sequenced RNA nucleotides. Translation is the process that converts the complementary genetic information carried by mRNA into the amino-acid sequence for the particular protein coded for in the DNA molecule. The two types of RNA that are employed in translation are mRNA and transfer RNA (tRNA).

b. The first step in protein synthesis is the binding of two subunits of a ribosome to a mRNA molecule. The initiation codon on the RNA molecule binds to the peptidyl site (P-site) on the ribosome. The mRNA/ribosome complex is now prepared for the first tRNA to arrive with its specific amino acid. After the anticodon of the tRNA aligns itself with the codon of the P-site, a second tRNA binds to the mRNA at the second or aminoacyl site (A-site) on the ribosome. The second tRNA aligns itself with its specific amino acid. Once the P-site and A-site are filled, an enzyme known as peptidyl transferase catalyzes the formation of a peptide bond between the two aligned amino acids.

c. Jacob and Monad conducted a series of bacterial experiments which provided them with information about gene regulation of protein synthesis in prokaryotes. From their data they developed two models of gene regulation. The first model begins with a regulatory gene coding for the synthesis of a repressor protein. Once synthesized, the repressor protein binds with an operator gene that is on another segment of the DNA molecule that coded for the synthesis of the repressor protein. The operator gene/repressor protein complex prevents transcription by the operator gene. If, however, the repressor protein combines with an inducer protein, the repressor protein cannot combine with the operator gene and therefore the operator gene is able to carry out transcription.

The second model proposed begins with a regulatory gene coding for the synthesis of a repressor protein. The repressor protein then binds with a corepressor protein instead of the operator gene and therefore allows the operator gene to carry out transcription. This suggests that either a structural gene is stimulated into transcription for an enzyme by the presence of a substrate for that enzyme or a structural gene is inhibited in transcription for an enzyme by the presence of a product resulting from the activity of that enzyme.

Question 2

a. (i) Allopatric speciation—Speciation that occurs when populations of a species live in different areas. Populations of the same species become geographically separated, which prevents their interbreeding. Each population is subject to different selective pressures and therefore becomes genetically different from the other. The test of speciation occurs when geographic barriers between the two supposed species break down and the two groups become sympatric. If they are truly different species, they will not be able to interbreed. New species of animal probably develop most often by means of allotropic speciation.
(ii) Sympatric speciation—Speciation that occurs within a single population in the same area. It usually occurs as a result of polypoidy. Two genetic mechanisms for instantaneous speciation are *autopolyploidy* (more than two chromosome sets of a species go into the formation of the zygote) and *amphiploidy* (diploid sets from two different species enter into the formation of the zygote). Possibly as many as one-third of all plant species may have arisen by means of polyploidy. Another sympatric speciation mechanism is *introgression*, successful hybridization leading to the incorporation of one species' genes into the gene complex of another species.
(iii) Parasympatric speciation—The evolution of reproductive isolation in two forms in adjacent populations. The reproductive isolation of the two forms results from contrasts in environmental factors that act as selective pressures favoring isolation of the two forms.

b. Prezygotic-isolating mechanisms are species-isolating phenomena that operate in preventing the zygote from forming.

c. (i) Habitat differences—Two species occupying separate habitats are kept apart and are therefore never in contact with one another for reproductive encounters.
(ii) Reproductive temporal differences—Two species of related animal that are capable of interbreeding if allowed to encounter one

another during times of reproductive maturity. Interbreeding never occurs because each species matures sexually at different times.
(iii) Mechanical differences—Two species in the same habitat are unable to carry out internal fertilization because they are physically incompatible: each species requires sexual organs of a specific size and shape in order to complete fertilization.
(iv) Behavioral differences—Two species carry on different courtship behaviors. Only members of the same species have the specific behavior patterns to complete the chain of events leading to sexual reproduction activities.

Question 3

The human respiratory system begins with the openings into the nose, the external nares. These lead to the nasal passages, which serve to warm and clean air that passes through them. The ends of the nasal passages open internally into the pharynx, by means of internal nares. The air passing through the pharynx moves through the glottis into the region of the larynx, or voicebox. The glottis is protected by a flap of skin that is known as the epiglottis. The epiglottis functions to prevent solid and liquid foods from entering the larynx instead of the esophagus. Air passing through the larynx enters the trachea, a thin-walled tube that is supported by c-shaped rings of cartilage that give the trachea its characteristic "lumpy" appearance. The trachea branches into a left and right bronchus. Each bronchus divides into a number of smaller bronchi until they become bronchioles that continue to subdivide into smaller and smaller bronchioles. The smallest brochioles open into alveolar ducts, each of which terminates in several alveolar sacs. Capillaries in the alveoli exchange gases with the air drawn into the lungs.

Inspiration and expiration in human external respiration requires changes in thoracic pressures. The thoracic cavity is a "closed system" that behaves in accordance with the following formula: $\mathbf{P \times V = K}$. Simply put, every time there is a volume change, there is an inverse pressure change. There are two ways in which thoracic volumes can be altered: (1) The diaphgram, a large dome-shaped muscle across the abdomen that serves as a barrier between the abdominal and thoracic cavities, can be contracted. The contraction of the diaphragm muscles causes it to straighten somewhat, thereby increasing the volume of the thoracic cavity. Relaxation of the diaphragm allows it to take on its characteristic dome-shaped structure and thereby reduces the volume of the thoracic cavity. (2) The intercostal muscles, those between the ribs, can be contracted to pull the ribs closer together and thereby pull the chest wall upward and outward. This action increases the volume of the thoracic cavity. Relaxation of the intercostal muscles allows the chest to return to its original position and thereby decreases the volume of the thoracic cavity.

An inspiration occurs in the following manner. Prior to any contraction in the muscles of respiration, the intrathoracic pressure is equal to the atmospheric pressure in the tubes and sacs of the lung. When muscles of respiration are contracted, they cause a decrease in the intrathoracic pressure. The atmospheric pressure therefore pushes more air into the lungs, causing them to inflate. This is known as negative pressure breathing. An expiration occurs when the muscles of respiration relax and cause an increase in the intrathoracic pressure. The increased intrathoracic pressure pushes against the outside of the lungs and causes the lungs to deflate.

Question 4

a. The current Fluid–Mosaic model of the cell membrane was proposed by Singer and Nicholson in 1972. This model pictures the cell membrane as a 7.5- to 10.0-nanometers thick structure composed of a bilayer of lipid molecules in which globular proteins are embedded. The molecules of lipid in the two layers are oriented with their nonpolar ends facing one another and their polar ends at the outer edges of the bilayer. Some cells exhibit an apparently orderly arrangement of the protein molecules embedded in the lipid layers, while other cells exhibit an apparent random arrangement. Protein and lipid composition and arrangement in the membrane appear to be related cell function. One important feature of the membrane structure is the relatively free movement of many protein and lipid molecules (fluidity) which permits alteration of the membrane's properties. Some membrane proteins appear to be attached to microtubules of the cell's endoskeleton and are therefore immobile. The basic fluid mosaic arrangement appears to be common to all membrane structures found in cells.

b. The cell membrane has the responsibility of regulating what gets into and out of the cell. Things enter and leave cells by means of simple or facilitated diffusion, osmosis, active transport, endocytosis, or by means of exocytosis. Movement across the cell membrane depends on a number of factors: lipid solubility, molecular size, molecular charge, and carrier molecules that require or do not require external energy.

In diffusion, molecules move along a concentration gradient across the membrane with no regard to the membrane. This probably occurs through pores in the membrane structure. Osmosis is a special case of diffusion in which water moves through the pores along a concentration gradient. Some molecules require carrier molecules within the membrane to facilitate movement along a concentration gradient. These carrier molecules do not require external sources of energy. There are carrier molecules that require energy to move molecules against concentration gradients across membranes. These carriers are involved in active transport mechanisms. Endocytosis is a mechanism for moving large particles into cells. This process employs part of the cell membrane surrounding the particle and forming a vacuole, or sac, that is brought into the cell through invagination. Exocytosis is essentially the reverse process: vacuole-surrounded materials are brought to the surface of the cell where the vacuole, or sac membrane, then fuses with the cell membrane, thereby releasing the contents to the outside of the cell.

APPENDIX

IMPORTANT NAMES IN BIOLOGY

Abel, John (1926)—Crystallization of insulin.

Allard, Howard (1920)—Description of the phenomenon of photoperiodism.

Banting, Frederick (1922)—Isolation of insulin. Nobel prize (1923) for work on insulin.

Beadle, George (1941)—Postulated the one gene–one protein hypothesis.

Best, Carl (1922)—See F. Banting.

Briggs, Robert (1953)—First successful nuclear transplants between frog zygotes.

Brown, Robert (1831)—Described the cell nucleus as an integral part of the cell.

Calvin, Melvin (1961)—Received a Nobel Prize for his work on the Dark Reaction chemistry in photosynthesis.

Chase, Martha (1950)—Demonstrated that the only part of a virus that enters a host cell is the DNA of the virus.

Cohen, Stanley (1973)—Succeeded in transplanting animal and plant genes into *E. coli.*

Correns, Karl (1900)—Rediscovered Mendel's laws simultaneously and independently along with H. DeVries and E. von Tschermak.

Crick, Francis (1953)—Described the structure of DNA, "the double helix." Nobel prize (1962).

De Duve, Christian (1961)—Described the cell organelle known as the lysosome.

De Vries, Hugo (1901)—Stated that mutations provide the variations observed among members of the same species. See also K. Correns.

Danielli, Joseph (1942)—Proposed the trilaminar sandwich model for the cytoplasmic membrane; protein–lipid–protein.

Darwin, Charles (1858)—Proposed the concept of evolution through natural selection. See A. Wallace.

Daveson, Henry—See J. Danielli.

Feulgen, Robert (1914)—Designed a procedure for the staining of DNA.

Fleming, Alexander (1929)—Discovered the antibiotic penicillin.

Flemming, Walther (1880)—Described the movement of chromosomes during mitosis.

Garner, Wayne (1920)—See H. Allard.

Golgi, Camillo (1898)—Described the reticular apparatus composed of folded sacs and vesicles composed of membranes now known as the Golgi Apparatus or Golgi Body.

Hardy, Gregory (1908)—Developed the Hardy–Weinberg Law which is used to provide a baseline against which real genetic populations can be compared to see if change is taking place.

Harvey, William (1628)—Demonstrated the circulation of the blood through the body with the heart being the pump for the system.

Hershey, Albert (1950)—See M. Chase.

Hooke, Robert (1665)—Coined the term "cell" to represent the compartments seen in cork slices.

Ingenhausz, Jan (1779)—Demonstrated that plants require sunlight for photosynthesis.

Jacob, Francois (1961)—Demonstrated that regulatory genes controlled genetic expression.

King, Thomas (1953)—See R. Briggs.

Koch, Robert (1882)—Established the germ theory of disease.

Kornberg, Arthur (1955)—Demonstrated that DNA could be replicated in a test tube environment.

Krebs, Hans (1953)—Received a Nobel Prize for his work in the elucidation of the tricarboxylic cycle in cellular respiration.

Landsteiner, Karl (1903)—Described the ABO blood system in humans.

Leder, Philip (1964)—Produced the synthetic RNA that led to the breaking of the DNA code.

Lamarck, Jean Baptist de (1809)—Proposed evolution by the inheritance of acquired characteristics.

Leeuwenhoek, Anton van (1673)—Using a microscope modified from the first microscope, designed by Jensen (1590), he described microorganisms in water.

Linnaeus, Carolus (1735)—Proposed the binomial system of classification for animals and plants.

Mangold, Hilde (1924)—Demonstrated the phenomenon of induction during embryonic development.

Mendel, Gregor (1866)—Published the first genetic studies and is now known as the "Father of Genetics."

Meselson, Matthew (1958)—Demonstrated that DNA replication is a semiconservative process.

Miescher, Friedrich (1869)—Demonstrated that the nucleus contained "nuclein," now termed "deoxyribonucleic acid" (DNA).

Miller, Stanley (1953)—Demonstrated that inorganic substances similar to those found in the primitive atmosphere of earth could form into simple organic compounds that serve as the foundation for life.

Mitchell, Peter (1978)—Received a Nobel Prize for developing the electron transport model for the production of ATP associated with the mitochondrial cristae (inner folded membrane).

Monod, Jacques (1961)—See F. Jacob.

Morgan, Lloyd C. (1893)—Enunciated Morgan's Canon which states that the interpretation of animal behavior should be done in the simplest possible neurological terms.

Morgan, H. Thomas (1940)—Postulated the existence of gene linkages.

Nicholson, Garth (1972)—Developed the fluid–mosaic model of the unit membrane for the cell.

Nirenberg, Marshall (1964)—See P. Leder.

Pasteur, Louis (1868)—Disproved the idea of spontaneous generation.

Pavlov, Ivan (1904)—Using dogs, he demonstrated the phenomenon of conditioned reflex.

Priestly, Joseph (1772)—Demonstrated that plants provided oxygen for animals.

Redi, Francesco (1665)—Demonstrated that "spontaneous generation" of maggots from decaying meat was the result of flies laying their eggs on the meat.

Robertson, David (1961)—Stated that all membranes in or around the cell had the same basic plan. He called it the "unit membrane."

Sanger, Fredrick (1954)—Determined the primary structure (amino acid sequence) of insulin. Nobel prize in 1958 for his work on insulin.

Schleiden, Matthis (1838)—Stated that all plants are composed of cells. Together with Schwann, they are responsible for the "Cell Theory."

Schwann, Theodore (1839)—Stated that all animals are composed of cells. Together with Schleiden, they are responsible for the "Cell Theory."

Singer, Stanley (1972)—See G. Nicholson.

Spermann, Hans (1924)—See H. Mangold.

Stahl, Franklin (1958)—See M. Meselson.

Stewart, Frederick (1951)—Demonstrated that every fully differentiated cell had all of the genes necessary to produce an entire new organism. Taking a specialized phloem cell from a carrot, he grew a complete carrot plant.

Tatum, Edward (1941)—See G. Beadle.

Urey, Harold (1953)—See S. Miller.

Virchow, Rudolph (1858)—Enunciated the concept of "Biogenesis," that is, living things come from pre-existing living things.

Van Baer, Karl E. (1828)–Proposed the germ layer theory to explain animal development.

Van Helmont, Jan Baptiste (1628)—Demonstrated that the soil was a minor contributor to the bulk of the plant body mass.

Van Tschermak, Erick (1900)—See K. Correns.

Wallace, Alfred (1858)—Independently arrived at the same conclusion about evolution that Darwin arrived at in the same year.

Watson, James (1953)—See F. Crick.

Weismann, August (1892)—Proposed the "germ plasma" theory of inheritance.

Wilkins, Maurice (1953)—See F. Crick.

Weinberg, Robert (1984)—Demonstrated that cancer is caused by "oncogenes."

Weinberg, Wendel (1908)—See G. Hardy.

Palade, George (1956)—Described the [illegible] for the cell.
Pasteur, Louis (1861)—See [illegible].
[illegible] (1877)—Discovered the [illegible].
[illegible] (1945)—[illegible] electron [illegible].
Priestley, Joseph (1772)—Demonstrated that plants produced oxygen in sunlight.
Redi, Francesco (1668)—Demonstrated that "spontaneous" generation of insects [illegible] was in fact the result of flies laying their eggs on it.
Robertson, David (1959)—[illegible].
Sanger, Frederick (1953)—Determined the primary structure [illegible] of insulin. Nobel Prize in 1958 for this work.
Schleiden, Mathias (1838)—Proposed that plants are composed of cells. [illegible].
Schwann, Theodor (1839)—Stated that all animals are composed of cells. Together with Schleiden, they are responsible for the cell theory.
Singer, Stanley (1972)—See Nicolson.
Spemann, Hans (1924)—See Mangold.
Stahl, Franklin (1958)—See Meselson.
Steward, Frederick (1958)—Demonstrated that every fully differentiated cell [illegible] the genes necessary to produce an entire new organism. Taking a single cell from a carrot, he grew a complete carrot plant.
Tatum, Edward (1941)—See Beadle.
Urey, Harold (1953)—See Miller.
Virchow, Rudolph (1855)—[illegible] cells come from pre-existing living cells.
Von Baer, Karl Ernst (1830)—[illegible].
Van Helmont, Jan Baptiste (1648)—Demonstrated that the soil was a minor contributor to the growth of the plant. [illegible]
van Leeuwenhoek, Anton (1676)—See [illegible].
Wallace, Alfred (1858)—Independently arrived at the theory of natural selection [illegible] the same year.
Watson, James (1953)—See Crick.
Weismann, August (1892)—Proposed the "germ plasm" theory of inheritance.
Wilkins, Maurice (1953)—See Crick.
[illegible] (1828)—[illegible].
[illegible]—See [illegible].

THE METRIC SYSTEM

Science uses the *metric system* of measurement. It is a simple system based on the number 10 and its multiples. Length is measured in *meters,* mass, or weight, is measured in *grams,* and volume is measured in *liters.* Specific multiples of 10 are designated by the following prefixes:

PREFIX	SYMBOL	MULTIPLE OF 10	
tera-	T	10^{12}	1,000,000,000,000
giga-	G	10^{9}	1,000,000,000
mega-	M	10^{6}	1,000,000
kilo-	k	10^{3}	1,000
hecto-	h	10^{2}	100
deka-	da	10^{1}	10
			UNIT (METER, LITER, or GRAM)
deci-	d	10^{-1}	0.1
centi-	c	10^{-2}	0.01
milli-	m	10^{-3}	0.001
micro-	μ	10^{-6}	0.000,001
nano-	n	10^{-9}	0.000,000,001
pico-	p	10^{-12}	0.000,000,000,001
femto-	f	10^{-15}	0.000,000,000,000,001
atto-	a	10^{-18}	0.000,000,000,000,000,001

The only other unit that should be mentioned is the Angstrom (Å), which is 10^{-10}m.

The metric measurement for temperature is the Celsius (formerly Centigrade) scale. On the Fahrenheit scale of temperature measure, water freezes at 32° and boils at 212°. On the Celsius scale water freezes at 0° and boils at 100°. The Celsius degree is therefore somewhat larger than the Fahrenheit degree. The conversion from one system to the other is simple:

$$\frac{[\text{Celsius degree}-32]}{1.8} = \text{Fahrenheit degree}$$

$$[\text{Fahrenheit degree} \times 1.8] + 32 = \text{Celsius degree}$$

MINIQUIZ ON THE METRIC SYSTEM

Try the following conversions.

1. 5 hectograms = ______ gram(s)
2. 19°C = ______ °F
3. 7 decigrams = ______ nanogram(s)
4. 9,779 liters = ______ dekaliter(s)
5. 7,691 liters = ______ microliter(s)
6. 49 decimeters = ______ meter(s)
7. 61°C = ______ °F
8. 9°F = ______ °C
9. 23 gigagrams = ______ megagram(s)
10. 81°C = ______ °F
11. 108 dekaliters = ______ liter(s)
12. 2 kiloliters = ______ liter(s)
13. 197 milligrams = ______ centigram(s)
14. 0.77 Angstroms = ______ nanogram(s)
15. 0.5 deciliters = ______ liter(s)
16. 37°C = ______ °F
17. 0.006 Angstom(s) = ______ millimeter(s)
18. 9 micrometer(s) = ______ nanometer(s)
19. 0.09 liter(s) = ______ microliter(s)
20. 119 deciliter(s) = ______ liter(s)

Answers to the Miniquiz

1. 500 grams
2. 66.2°F
3. 70,000,000 nanograms
4. 97,790 dekaliters
5. 7,691,000 microliters
6. 4.9 meters
7. 141.8°F
8. −12.8°C
9. 23,000 megagrams
10. 177.8°F
11. 10.8 liters
12. 2000 liters
13. 19.7 centigrams
14. 7.7 nanograms
15. 0.05 liters
16. 98.6°F
17. 0.0006 milimeters
18. 9000 nanometers
19. 90,000 microliters
20. 11.9 liters

SOME CLASSIC EXPERIMENTS IN BIOLOGY

It is recommended that the following list of experiments in biology be reviewed. This book does not have the space to include the specific descriptions of each of the experiments, but they are described in detail in many college biology textbooks.

Jan Baptista van Helmont (1630s) Experiments in support of the idea that soil plays a minor role in the contribution of nutritional material for growth in a plant

Joseph Priestly (1770s) Experiments in support of the idea that gases play an important role in the growth of green plants

Gregor Mendel (1850s) Experiments in support of the idea that inherited characteristics are carried by discrete units

Louis Pasteur (1860s) Experiments that disproved the concept of spontaneous generation of animals and plants

Robert Koch (1880s) Experiments in support of the idea that some diseases are caused by microscopic organisms

Thomas Hunt Morgan (1910s) Experiments in support of the idea that certain genes are sex-linked

Francis Griffith (1920s) Experiments in support of the idea that there is a transforming factor in bacteria

Fritz Went (1920s) Experiments in support of the idea that shoot tips produce substances that control cell elongation in developing stems

Joachim Hämmerling (1930s) Experiments in support of the idea that the cell nucleus controls the cell cytoplasm

O. T. Avery, C. M. McLoed, and M. McCarty (1940s) Experiments in support of the idea that the transforming factor in bacteria is DNA

George Beadle and Edward Tatum (1940s) Experiments that contribute to the understanding of the mechanism of gene expression

Stanley Miller (1950s) Experiments in support of the idea that a primitive atmosphere of several specific gases is capable of giving rise to simple organic molecules

Alfred Hershey and Martha Chase (1950s) Experiments in support of the idea that the genetic material in viruses is DNA and not protein

Frederick Stewart (1950s) Experiments in support of the idea that a specialized plant cell has all of the genes to develop into a complete plant

W. W. Garner and H. A. Allard (1950s) Experiments in support of the idea that plants respond to the length of light and dark periods

Sidney Fox (1960s) Experiments in support of the idea that simple organic molecules are capable of forming proteinoid microspheres that can mimic the activities of simple cells

Matthew Meselson and Franklin Stahl (1960s) Experiments in support of the idea that DNA replicates by means of a semiconservative mechanism

Joseph McDade and Charles Shepard (1970s) Experiments that identified the organism responsible for Legionnaires' disease

TEXTBOOK REFERENCES

Armes, Karen, and Pamela S. Camp. *Biology*, 3rd Edition. Chicago, London, New York, Philadelphia, San Francisco: Saunders College Publishing, 1987.

Barrett, James M., Peter Abramoff, A. Krishna Kumaran, and William F. Millington. *Biology*, 1st edition. Englewood Cliffs, New Jersey: Prentice Hall, 1986.

Kimball, John W., *Biology*, 5th edition. Reading, MA and Menlo Park, CA: Addison Wesley, 1983.

Wessells, Norman K., and Janet L. Hopsen. *Biology*, 1st Edition. New York: Random House, 1988.

Audersirk, Gerald, and Teresa Audersirk. *Biology—Life on Earth*. New York: Macmillan, 1989.

Starr, Cecie, and Ralph Taggart. *Biology—The Unity and Diversity of Life*. Belmont, CA: Wadsworth, 1989.

Keeton, William T., and James L. Gould. *Biological Sciences*, 4th Edition. New York and London: W. W. Norton, 1986.

Mader, Sylvia S. *Inquiry into Life*, 5th Edition. Dubuque, IA: William C. Brown, 1988.

Curtis, Helena, and N. Sue Barnes. *Invitation to Biology*, 4th Edition. New York: Worth Publishers, 1985.

Purves, William K., and Gordon H. Orians. *Life—The Science of Biology*, 2nd Edition. Sunderland, MA: Sinauer Associates, 1987.

TEXTBOOK REFERENCES

Arms, Karen, and Patricia S. Camp. *Biology*, 3rd Edition. Chicago, London, New York, Philadelphia and San Francisco: Saunders College Publishing, 1987.

Barrett, James M., Peter Abramoff, A. Krishna Kumaran, and William F. Millington. *Biology*, 1st edition. Englewood Cliffs, New Jersey: Prentice Hall, 1986.

Kimball, John W., *Biology*, 5th edition. Reading, MA and Menlo Park, CA: Addison Wesley, 1983.

Wessells, Norman K., and Janet L. Hopson. *Biology*, 1st Edition. New York: Random House, 1988.

Audesirk, Gerald, and Teresa Audesirk. *Biology—Life on Earth*. New York: Macmillan, 1986.

Starr, Cecie, and Ralph Taggart. *Biology—The Unity and Diversity of Life*. Belmont, CA: Wadsworth, 1989.

Keeton, William T., and James L. Gould. *Biological Science*, 4th Edition. New York and London: W. W. Norton, 1986.

Mader, Sylvia S. *Inquiry into Life*, 5th Edition. Dubuque, IA: William C. Brown, 1988.

Curtis, Helena, and N. Sue Barnes. *Invitation to Biology*, 4th Edition. New York: Worth Publishers, 1985.

Purves, William K., and Gordon H. Orians. *Life—The Science of Biology*, 2nd Edition. Sunderland, MA: Sinauer Associates, 1987.